# Waste Treatment

## REDUCING GLOBAL WASTE

# GREEN TECHNOLOGY

# *Waste Treatment*

## *REDUCING GLOBAL WASTE*

Anne Maczulak, Ph.D.

☑ Facts On File

*An imprint of Infobase Publishing*

**WASTE TREATMENT: Reducing Global Waste**

Facts On File, Inc.
An imprint of Infobase Publishing
132 West 31st Street
New York NY 10001

**Library of Congress Cataloging-in-Publication Data**

Maczulak, Anne E. (Anne Elizabeth), 1954–
  Waste treatment : reducing global waste / Anne E. Maczulak.
      p. cm. — (Green technology ; v. 2)
  Includes bibliographical references and index.
  ISBN-13: 978-0-8160-7204-0
  ISBN-10: 0-8160-7204-3
  1. Refuse and refuse disposal—Juvenile literature. 2. Waste minimization—Juvenile literature. 3. Recycling (Waste, etc.)—Juvenile literature. I. Title.
  TD792.M35 2010
  628.4—dc22      2008045054

Facts On File books are available at special discounts when purchased in bulk quantities for businesses, associations, institutions, or sales promotions. Please call our Special Sales Department in New York at (212) 967-8800 or (800) 322-8755.

You can find Facts On File on the World Wide Web at http://www.factsonfile.com

Text design by James Scotto-Lavino
Illustrations by Bobbi McCutcheon
Photo research by Elizabeth H. Oakes

Printed in the United States of America

Bang Hermitage 10 9 8 7 6 5 4 3

This book is printed on acid-free paper.

# Contents

# Preface

The first Earth Day took place on April 22, 1970, and occurred mainly because a handful of farsighted people understood the damage being inflicted daily on the environment. They understood also that natural resources do not last forever. An increasing rate of environmental disasters, hazardous waste spills, and wholesale destruction of forests, clean water, and other resources convinced Earth Day's founders that saving the environment would require a determined effort from scientists and nonscientists alike. Environmental science thus traces its birth to the early 1970s.

Environmental scientists at first had a hard time convincing the world of oncoming calamity. Small daily changes to the environment are more difficult to see than single explosive events. As it happened the environment was being assaulted by both small damages and huge disasters. The public and its leaders could not ignore festering waste dumps, illnesses caused by pollution, or stretches of land no longer able to sustain life. Environmental laws began to take shape in the decade following the first Earth Day. With them, environmental science grew from a curiosity to a specialty taught in hundreds of universities.

The condition of the environment is constantly changing, but almost all scientists now agree it is not changing for the good. They agree on one other thing as well: Human activities are the major reason for the incredible harm dealt to the environment in the last 100 years. Some of these changes cannot be reversed. Environmental scientists therefore split their energies in addressing three aspects of ecology: cleaning up the damage already done to the Earth, changing current uses of natural resources, and developing new technologies to conserve Earth's remaining natural resources. These objectives are part of the green movement. When new technologies are invented to fulfill the objectives, they can collectively be called green technology. Green Technology is a multivolume set that explores new methods for repairing and restoring the environment. The set covers a broad range of subjects as indicated by the following titles of each book:

- *Cleaning Up the Environment*
- *Waste Treatment*
- *Biodiversity*
- *Conservation*
- *Pollution*
- *Sustainability*
- *Environmental Engineering*
- *Renewable Energy*

Each volume gives brief historical background on the subject and current technologies. New technologies in environmental science are the focus of the remainder of each volume. Some green technologies are more theoretical than real, and their use is far in the future. Other green technologies have moved into the mainstream of life in this country. Recycling, alternative energies, energy-efficient buildings, and biotechnology are examples of green technologies in use today.

This set of books does not ignore the importance of local efforts by ordinary citizens to preserve the environment. It explains also the role played by large international organizations in getting different countries and cultures to find common ground for using natural resources. Green Technology is therefore part science and part social study. As a biologist, I am encouraged by the innovative science that is directed toward rescuing the environment from further damage. One goal of this set is to explain the scientific opportunities available for students in environmental studies. I am also encouraged by the dedication of environmental organizations, but I recognize the challenges that must still be overcome to halt further destruction of the environment. Readers of this book will also identify many challenges of technology and within society for preserving Earth. Perhaps this book will give students inspiration to put their unique talents toward cleaning up the environment.

# Acknowledgments

I would like to thank a group of people who made this book possible. Appreciation goes to Bobbi McCutcheon who helped turn my unrefined and theoretical ideas into clear, straightforward illustrations. Thanks also go to Elizabeth Oakes, Ph.D., for providing photographs that recount the past and the present of environmental technology. My thanks also go to Marilyn Makepeace, who provided support and balance to my writing life, and Jodie Rhodes, who helped me overcome more than one challenge. I appreciate the guidance I received from Bruce J. Murphy of IntelliWaste, Inc., on the fine points of waste management and from S. K. Sundaram, Ph.D., of the Pacific Northwest National Laboratory on the background of vitrification. Finally, I thank Frank Darmstadt, executive editor, for his patience and encouragement throughout my early and late struggles to produce a worthy product. General thanks go to Facts On File for giving me this opportunity.

# Introduction

Every living thing, from microscopic bacterial cells to giant redwood trees, takes in nutrients and excretes wastes. Nutrients fuel the inner workings of all animals, plants, and single-celled organisms. After using nutrients, each cell of every living thing produces waste. Biological wastes from one organism are very often used as nutrients by another being. An easy example to visualize is the oxygen given off as an end product of photosynthesis in plants, which is then used by animal cells. This form of recycling serves a useful purpose, because if biological end products accumulate in the environment, they eventually inhibit other forms of life.

Humans have developed most of their working machinery based on the simple biological model of nutrients in and wastes out. Humans take in fuel and expel wastes; machines take in fuel and expel wastes. Wastes from equipment, vehicles, appliances, and other nonbiological things would build up and halt human activities if they were left unattended, just the same way excess cellular end products begin to harm cells. The main end products from machinery are gas *emissions,* used oil, ash, and heat. The subject of this book is waste treatment technology. Waste treatment is the removal of wastes from the environment by burning, decomposing, or chemically transforming them so that Earth's activities can continue. It is one of the most critical phases of *waste management.*

This volume in the Green Technology set explores how the waste treatment industry plays a role in removing, treating, and disposing of human, household, and industrial wastes. *Waste Treatment* begins with a look at the global waste problem. It defines the different classifications of materials that are treated today in waste management. One of the most important concepts in waste management is the *waste stream*. Waste streams are all the sources of various wastes as they move through the environment toward a final disposal. The control of waste streams is the central theme throughout this book.

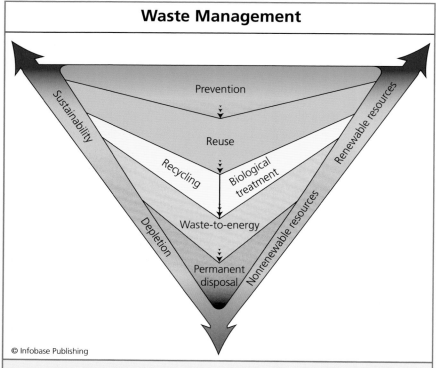

## Waste Management

Modern waste management programs aim to minimize the total amount of nonreusable waste and maximize the amount of reusable waste. The ultimate goal in sustainability is waste prevention.

This book also presents the ways in which *hazardous* and *nonhazardous wastes* are defined. These classifications are more than just a curiosity. Waste managers can make better decisions on treatment methods when wastes are grouped by their physical form, chemical content, degree of hazard to the environment, or source. These groupings also allow environmental scientists to learn about the trends in our society's waste and in society itself. Waste types can change quite dramatically in a period of less than 100 years. For example, this book shows how wastes from electronic products are a big problem in the world today but certainly were not a concern in the early 1900s. But in 1910, for instance, horse manure from the use of thousands of horse-drawn vehicles was probably a huge waste problem!

Chapter 1 gives an overview of the world's waste problem. Special aspects of waste management are explained. Some of the key aspects are the following: (1) the reasons for waste categories; (2) selection of the best waste treatment method; (3) the role of landfills; (4) waste transport requirements; and (5) the relationship between waste buildup and *ecosystem* health.

Chapter 2 discusses the unique challenge of one of the world's most pressing waste problems: discarded electronic products. These items have accumulated quickly in the past two decades. Stockpiles of used or obsolete electronic waste, *e-waste,* are reaching alarming levels in developed and developing countries. The chapter discusses why e-waste is a particular hazard in developing regions of the world. The treatment of e-waste is unlike that of any other waste. The chapter describes the steps for *salvaging* the components of e-waste and the special hazards contained in this waste category.

Chapters 3 and 4 present the advantages and disadvantages of two thermal methods in waste treatment: *incineration* and *vitrification.* Incineration has been a dependable waste treatment method for a century. Chapter 3 discusses familiar drawbacks and perceptions of incinerator emissions. It also describes new technologies for changing incineration from an undesirable treatment method to a surprisingly groundbreaking technology and offers a case study on the development of the Clean Air Act.

Chapter 4 focuses on the world's most innovative thermal method for treating highly hazardous radioactive and nonradioactive wastes: vitrification. It describes this technology and the reason why it may become the United States's last best hope for disposing of its stockpile of nuclear wastes. It also illustrates the hurdles towns face when they desire new waste treatment technologies. Finally, the chapter explains the basics of radioactive materials.

Chapter 5 looks at ways in which wastes in the environment can be made stationary so they do not harm uncontaminated places. *Solidification* and *stabilization* name two related technologies that are now using new chemical formulas and simple biological techniques to hold pollutants in the soil in a safe form. The chapter pays special attention to the current status of Yucca Mountain's proposed hazardous waste site. The problems related to this government operation are described in this chapter, but Yucca Mountain is mentioned throughout the book because of its importance to a number of hazardous waste programs in this country.

Chapter 6 reviews two technologies used on wastes that are not treated by traditional *combustion* methods. It describes reduction methods and

| Pollution Cleanup/Treatment Technology | | |
|---|---|---|
| Cleanup | Combination | Treatment |
| Excavation | Bioremediation | Incineration |
| Extraction | Solidification | Vitrification |
| Physical separation | Stabilization | Biodegradation |
| Steam stripping | Oxidation–Reduction | Encapsulation |
| Smelting | Neutralization | Wastewater treatment |
| Adsorption | | Salvage/Recycling |
| © Infobase Publishing | | Landfill/Burial |

Waste management has increasingly turned to methods that combine pollution cleanup with treatment. Other technologies dedicated to either cleanup or treatment may then supplement cleanup/treatment combinations. This increases the overall efficiency of managing hazardous wastes.

compaction methods. Special attention is given to the ways in which compacted materials are now designed for sustainable uses.

*Waste Treatment* concludes with a chapter describing wastewater treatment. The chapter explains how wastewater treatment is actually a form of *bioremediation* and why wastewaters are distinct from almost every other kind of hazardous waste. The current chemical, physical, and biological steps in wastewater treatment are covered as well in new technologies for removing biological and nonbiological waste matter. This chapter also delves into approaches for using wetlands to help purify surface waters.

*Waste Treatment* follows a theme begun in *Cleaning Up the Environment,* the first book in the Green Technology set. Today's hazardous waste management is usually a mixture of cleanup and treatment methods at the same hazardous waste site. Hazardous waste stockpiles are also increasingly being managed with technologies that combine cleanup and treatment within the same process. In fact, few projects in contamination cleanup do not use some combination of methods. Cleaning Up the Environment and Waste Treatment describe these cleanup/treatment technologies and why they are an advantage in hazardous waste management.

Perhaps the most interesting message offered by these chapters is the relationship between society and its wastes. The types, amounts, and storage of wastes in the world today tell a story about the way people live. They give clues about society's level of technology. Waste buildup or its reduction over time also tells scientists how well populations are doing in restoring their planet.

# ASSESSING GLOBAL WASTE MANAGEMENT

A typical person living in an industrialized country discards about 4.5 pounds (2 kg) of solid waste each day, but household *garbage* makes up only a portion of the solid wastes generated every day. Offices, construction sites, restaurants, farms, and manufacturing plants produce most of the solid wastes generated daily. In addition to solid wastes, thousands of gallons of wastewaters from towns and cities and hazardous liquids from businesses contribute to total global waste. Before any community, city, or country can safely remove and dispose of these materials, people must understand the nature of waste, meaning its solid or liquid characteristics and its potential hazards. Waste management comprises all activities that deal with every aspect of solid and liquid waste: collection, transport, recycling, and disposal.

At present there is hardly a place on Earth that has not been exposed to some sort of waste. Some of these materials cause immediate health hazards to humans and animals. Other wastes persist for years in the environment until they reach levels damaging to healthy ecosystems. An ecosystem is the complex of plants and animals that interact with each other and their surrounding environment. It is critical to keep ecosystems working properly because the health of Earth's *biomes* depends on the combined activities of individual ecosystems. *Pollution* causes a situation in ecosystems called ecosystem imbalance in which food and physical conditions are no longer adequate for the ecosystem's normal inhabitants. Damaged ecosystems soon disrupt the normal workings of entire communities, which are all the populations of living things in a defined area.

In time, Earth's biomes feel the effects of waste buildup. Waste managers today know that even small amounts of waste may in time lead to global environmental problems. This chapter describes the types of waste that upset ecosystems and the ways in which small amounts of waste can grow into large environmental hazards. The chapter introduces the concept of waste streams, describes the responsibilities in waste management, and discusses two important aspects of waste management: transport and landfill disposal. Finally, this chapter describes the increasingly important salvage industry.

## THE GROWING PROBLEM OF GARBAGE AND WASTE

People know waste when they see it. A Dumpster piled high with garbage bags, a pickup filled with old computers, containers of used aluminum cans and newspapers—these are obvious signs of waste. Additional materials enter the environment each day less noticed. These unseen materials are of greatest concern because they enter ecosystems silently. They may be chemicals dissolved in river water, gases in car emissions, or tiny bits of oil in dunes on a beach. In order to understand the total amount of wastes entering the environment, all of the visible and invisible substances must be considered.

The waste materials made in any region of the world can be thought of as related to the population's wealth, because wealth often affects a region's technologies. Industrialized countries annually generate more than 450 million tons (408 million metric tons) of solid waste. In the United States alone, solid waste generation has increased 235 percent in the last 40 years to more than 12 billion tons (11 billion metric tons) annually. Despite this growth, disposal methods remain quite primitive, especially when compared with advances in other technologies—computers, space exploration, and biotechnology—during the same period. Burning and burying still play major roles in waste disposal as they did in the earliest human societies.

The types of waste have changed throughout human history as technology has changed, but the puzzle of how to dispose of them has lasted. Apparently early civilization had as difficult a time in waste management as people do today. For instance, archaeologists examining ancient sites dating to 6500 B.C.E. in what is now Colorado have determined that settlement dwellers may have discarded as much as five pounds (2.3 kg) of waste a day.

The history of waste began with the history of humans, but waste grew into a serious problem when societies began building their commerce. For centuries, people living in rural areas or towns either burned much of their household waste or dumped it into swamps and rivers. Waterways became so clogged in England that Parliament in 1388 banned the use of rivers for waste disposal simply so boats could make their way upstream. The land took its share of wastes too. In 1400, the garbage hauled out of Paris formed mountains so great outside the city that travelers were hard-pressed to find routes in and out.

The United States experienced a similar dilemma when its population expanded and its economy began to grow. U.S. businesses followed the familiar prescriptions for waste disposal: burning, burying, or dumping into waters. In time, U.S. waterways had become almost as clogged as the English rivers in the Middle Ages, and by 1899 Congress passed the Rivers and Harbors Act to ban the discharge of solid and liquid wastes into waterways used by boats. Despite these steps, people remained surprisingly slow to grasp the dangers of toxic and infectious materials filling the environment. Not until 1978 when the Love Canal area near Niagara Falls, New York, became so engorged with dumped chemicals that they seeped into schools and homes and made residents ill, did the government awaken to the need for hazardous waste controls. Today, individual and industrial wastes are managed more carefully than in the past, although waste-disposal innovations have been slow to emerge.

Waste management today can be divided into two major areas of emphasis: (1) the reduction of waste production at its source and (2) the development of better technologies for treating waste. It all begins with knowing as much as possible about the composition of waste.

## HAZARDOUS AND NONHAZARDOUS WASTE

Solid and liquid wastes are of two types: hazardous and nonhazardous. Hazardous wastes consist of liquids, solids, or gases that are toxic or corrosive or can ignite or react in the air or with other chemicals. Biohazards are pathogenic (disease-causing) microbes, used needles and bandages, and blood and other bodily fluids, and all of these are considered an infectious form of hazardous waste because they might transmit disease. The U.S. Environmental Protection Agency (EPA) includes a subcategory of hazardous wastes called *universal wastes.* These substances do not meet the definitions given here, but they can be a hazard in the environment. Items within the classification of universal wastes are the following: batteries, pesticides, fluorescent bulbs, mercury-containing thermometers, and other equipment with toxic metals.

The EPA is responsible for enforcing the laws controlling hazardous waste in the United States, and it groups waste by three main methods: (1) chemical composition, (2) source, or (3) industry. When classifying wastes by chemical composition, the EPA and the waste industry further categorize the substances into the following groups, each of which have their own subgroups: chlorinated organic compounds, mercury-containing chemicals, military munitions, paint-manufacturing wastes, phenols, and *radioactive wastes.* Many of the most hazardous waste products from the home (paints, mercury-containing thermometers, motor oil, antifreeze, solvents, and chemical pesticides) often fit into these same categories. Another aspect of waste complicates any classification system: A large number of waste substances can belong to more than one category.

The EPA also oversees the handling of nonhazardous wastes. Though these materials are not toxic, they can fill up habitats and interfere with ecosystems if they are left unattended. Nonhazardous wastes are paper, packaging, plastic, nontoxic metals, glass, yard trimmings, wood chips, and construction waste. *Municipal solid waste* (MSW) also contributes to the total tonnage of nonhazardous waste. MSW contains garbage from households and businesses plus yard trimmings, wood, glass, small appliances, clothing, and pieces of furniture. The waste management industry strives to monitor MSW so that it contains only materials that do not cause harm to human or animal health or have toxic effects on the environment.

The EPA's classification of wastes based on source often gives an approximate idea of its composition as shown in the following table. These categories help waste managers speculate on the waste's general composition, but they do not provide enough information to define exact composition. For example, a waste manager would have a fairly good idea of the chemicals in mining wastes but would not be able to predict the day-to-day components of agricultural wastes.

| TYPES OF WASTE | |
|---|---|
| **TYPE OR SOURCE** | **DESCRIPTION OF CONTENTS** |
| municipal solid waste (MSW) | household, hotels/motel, and business trash and garbage: food scraps, bottles, packaging, paper, newspapers, batteries, yard trimmings, furniture, appliances, clothing, and toys |
| EPA-regulated hazardous waste | hazardous substances monitored by the EPA by law: substances that are ignitable, corrosive, reactive, toxic, or etiologic |
| radioactive waste | any solid, semisolid, or liquid waste containing radioactive elements |
| wastes from extraction industries | wastes from mining and mineral processing: metals, minerals, acids, and solvents |
| industrial nonhazardous waste | excess materials from manufacturing or energy production: pulp and paper, iron and steel, glass, plastics, and concrete |
| household hazardous waste | household items containing EPA-regulated chemicals: paints, stains, varnishes, solvents, cleaning chemicals, and pesticides |
| agricultural waste | animal waste from livestock, dairies, other farm animals and wastes from crop production and harvesting: manure, feed, used bedding, carcasses, and crop discards such as leaves, vines, twigs, branches, and weeds |
| construction/ demolition waste | debris from construction, renovations, remodeling, or demolitions: wood, concrete, brick, steel and other metals, glass, drywall, plaster, and insulation |
| medical waste | solids generated in diagnosis, treatment, or immunization of humans or animals and from clinical, research, or manufacturing settings: unused drugs, needles, syringes, bottles and tubing, bandages, wraps, bedding, medical and dental devices, and protective clothing |

*(continues)*

| TYPES OF WASTE *(continued)* | |
|---|---|
| **TYPE OR SOURCE** | **DESCRIPTION OF CONTENTS** |
| oil and gas industry waste | solids and liquids produced in exploration, drilling, and production of crude oil or natural gas |
| *sludge* | solid, semisolid, or liquids from wastewater treatment |
| dredging waste | solids and semisolids removed from the bottom of rivers and harbors |
| *sewage* | household or industrial wastewaters discharged into sewers |

Wastes that emerge each day from cities, households, and factories do not fit into exact categories bases on composition because any waste load's components vary from one load to the next. Waste typically contains a mixture of hazardous and nonhazardous substances and an assortment of chemical and biological matter. For example, a discarded electronic device contains nontoxic plastics and metals that make up the outer shell, but it also holds toxic lead, mercury, and cadmium. Similarly, a bag of medical waste likely holds infectious microbes and blood, mercury compounds, cleaning solvents, and perhaps radioactive matter in addition to less dangerous items. Even if a waste load has been identified as hazardous, there may be a complex mix of hazardous substances in that one load, even newer chemicals that did not exist even a few years before. In 1980, *Time* magazine correspondent Ed Magnuson noted, "Of all man's interventions in the natural order, none is accelerating quite so alarmingly as the creation of chemical compounds."

In addition to a waste load's composition, the waste treatment industry considers the source of each material that requires treatment. This information helps waste managers develop better ways of *sorting* wastes so that hazardous materials receive the correct treatment method and nonhazardous wastes follow their own path to disposal. By knowing the composition as well as the source of a waste load, waste treatment facilities can predict how quickly the materials will decompose. Materials that decompose quickly are treated differently than matter that persists for years, perhaps thousands of years. For instance bacterial toxins are lethal, yet they break down readily in normal wastewater treatment. *Radioisotopes* from nuclear reactors on the other hand can persist for hundreds of thousands of years.

Modern waste management includes more responsibilities than simply dispatching a fleet of garbage trucks. Part of its job now is to consider natural resource use and *sustainability,* to concentrate on making the greatest use of all components of each load for the purpose of reducing the unusable portion that goes to disposal. By reusing materials in waste, society conserves many natural resources. Waste management is, for this reason, an important part of green technology. Using less of the world's natural resources reduces total waste output. Sustainable waste management follows a three-pronged approach to not only reduce waste but to reduce its effect on the environment and to possibly get a benefit from certain wastes. These three complementary approaches are: (1) safe and efficient handling of waste; (2) programs for reducing waste generation; and (3) recycling technology.

Sustainable waste management begins by dividing waste into two groups: preconsumer and postconsumer. *Pre-consumer waste* consists of the leftover materials generated in the manufacture of products. In many instances, it is recycled at the manufacturing plant so that the plant produces a smaller final load. But even at top efficiency certain industries produce a lot of waste. For example, oil and gas production and the mining industry generate more than half of the solid waste produced in the United States each year. The EPA classifies oil and gas production wastes as "wastes generated during the exploration, development, and production of crude oil, natural gas, and geothermal energy." Mining, especially mountaintop mining in which equipment slices away an entire mountain peak, creates its own unique waste problem. *New York Times* correspondent John Broder explained in a 2007 article on mountaintop mining, "All mining generates huge volumes of waste, known as excess spoil or overburden, and it has to go somewhere." Industries such as pulp and paper, metal, and agriculture also produce large amounts of solid waste. The food industry—food product companies and meat production—contributes to the total of nonhazardous pre-consumer waste.

*Postconsumer waste* consists of unused materials and packaging left after consumers purchase and use products. Packaging makes up a large proportion of postconsumer waste and much of it goes into municipal recycling programs. As sustainable waste management improves in the future, companies will be expected to reduce the amount of packaging they use in order to reduce postconsumer waste.

Industry has an important responsibility in reducing total waste produced because pre-consumer waste has been estimated as 25 times that of postconsumer waste. Unfortunately, community recycling programs that handle postconsumer waste have been more successful than many industry recycling initiatives that would have an impact on pre-consumer wastes. This is because industries make economic decisions on recycling; recycling has a better chance of succeeding in industry if it helps save money. Currently, the textiles and carpet industries and paper manufacturers produce a large proportion of this country's pre-consumer waste.

Recyclers help sustainable waste management by knowing the potential value of materials found in waste loads. The main uses for recycled waste are as raw materials for new products or as a fuel for energy production. Materials most useful for recycling are rubber, plastic, aluminum and other metals, glass, paper, and wood. The amounts of these wastes

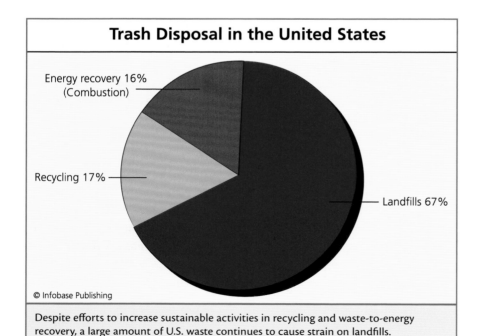

**Trash Disposal in the United States**

Energy recovery 16% (Combustion)

Recycling 17%

Landfills 67%

© Infobase Publishing

Despite efforts to increase sustainable activities in recycling and waste-to-energy recovery, a large amount of U.S. waste continues to cause strain on landfills.

stagger the imagination: The United States discards enough aluminum each year to rebuild its entire fleet of airplanes. In order to make better use of these waste materials, more people must think of waste as a valuable raw material for new products rather than another load for the landfill. Communities can start by studying many examples of small companies that make new products from waste materials. Rubbersidewalks, Inc., is a California company that converts discarded tires into pedestrian sidewalks. The innovative surfaces require fewer repairs and allow easier tree and root maintenance than more costly concrete sidewalks. A 2006 EPA press release described the environmental value of this invention: "Finding a new use for old tires is important because piles of scrap tires can become breeding grounds for disease-carrying pests such as mosquitoes. In addition, tire pile fires are difficult to extinguish and release smoke that is dangerous to both human health and the environment. The new sidewalks not only use old and unwanted tires, but they can also help save urban trees. Traditional concrete sidewalks conflict with tree growth by cutting off the roots' air and water supply."

## WASTE STREAMS

A waste stream is the waste output of a community, region, or state and the manner in which it moves to a final disposal site. Wastes come from many directions and sources: people, farms, manufacturing plants, office buildings, households, and nature. Either as solids or liquids, these materials follow a variety of routes toward specific disposal sites: recycling centers, landfills, incineration plants, or sewage treatment plants.

Each waste stream starts at a source. Sources of waste range from a small rodent in a meadow to a massive manufacturing plant. A single geographic region or even a small community contains numerous routes through which waste flows: rivers, streams, storm drains, sewers, Dumpsters, garbage cans, or smokestacks. Neighborhoods have their own characteristic waste streams that differ from the waste streams of larger towns and cities, which differ from agricultural regions or recreational areas. A single neighborhood block may contain many places where waste starts out, beginning for instance with houses, one or more office buildings, a doctor or dentist's office, a hair salon, restaurants, a park, a cleaners, and a copy shop. All of these establishments produce garbage containing food, paper, electronic devices, excess furniture, clothing, etc., all the

## Flow of Toxic Substances

© Infobase Publishing

The environment receives a diversity of hazardous materials every day. A large portion of these materials flows with rain runoff or in surface waters toward large rivers, lakes, and the ocean. Toxic substances that accumulate in aquatic ecosystems cause serious harm to food webs and biodiversity.

components that make up MSW. After items are discarded, they take different routes to a final disposal site. Garbage trucks pick up loads and haul them away, while lawn trimmings, pesticides sprayed on fruit trees, and animal wastes may wash into storm drains. Meanwhile, spilled gasoline, oil leaks, and car emissions add to the air, land, and water waste streams. These limited examples illustrate the enormous variety of waste streams and sources that contribute to an area's daily waste total.

Waste streams are best understood by thinking of an uncomplicated example. A toilet flushes, the material moves in pipes to a wastewater treatment plant where biological (microbes) and chemical (particles that cause settling) activities remove hazardous components. The treated water is reused for irrigation, sent to industrial processes, or released into a nearby body of water. A more complicated waste stream consists of many more routes and wastes that contain different substances all mixed together. In a perfect world, even complicated waste streams are controlled until treatment facilities remove all hazards and prevent them from reaching the environment. This is not a perfect world. People sometimes interfere with efficient waste streams by demanding that a landfill near their neighborhood be closed or a nuclear waste site be banned in their state. Even garbage hauler strikes stall waste streams and cause toxic materials to build up. When waste streams are disrupted, hazardous substances never reach their intended treatment site and instead contaminate soil and water, which then damages ecosystems.

Disrupted nonhazardous waste streams can also damage the environment even though they contain no hazardous materials. For example, when garbage litters beaches or parks, it poses a danger to wildlife. Discarded fishing lines tangle birds' bills and bind the mouths of marine mammals, causing them to starve. Small bits of swallowed glass or foil wrappers damage digestive tracts. Tires may block a lagoon's flow and affect aquatic life in the nearby wetlands. These examples represent a small sampling of the many human activities that affect waste streams every day.

## ECOSYSTEM HEALTH

The Earth's ecosystems play a vital role in recycling nutrients. *Nutrient cycling* refers to the transformation of elements in nature from organic form to inorganic form and back again. Carbon, nitrogen, phosphorus, sulfur, potassium, minerals, and water all have their own cycles, also called *biogeochemical cycles*. In a nutrient cycle, an element or a molecule leaves a body when an animal or plant dies and decomposes. This element may then enter the atmosphere, return to the earth, be consumed as a nutrient by a living thing, and then return to the earth when the living entity dies. Ecosystem *food webs* contribute to nutrient recycling because nutrients move through food webs by way of a variety of single-celled and multicellular organisms. In a single nutrient cycle an element may become part of various chemical forms, in many different organisms from bacteria to large mammals. In the carbon cycle, for example, carbon takes the form of a gas, an insoluble solid, and a water-soluble compound all within one cycle, detailed in the following table.

Wastes that kill microbes or animals or stunt the growth of plants upset the carbon cycle. It is easy to imagine similar damage done by wastes to the cycling of nitrogen, sulfur, and the other nutrients. When

| CARBON FORMS AS IT CYCLES IN THE ENVIRONMENT | |
|---|---|
| **PHASE OF THE CARBON CYCLE** | **CARBON'S FORM** |
| atmosphere | carbon dioxide gas |
| photosynthesis in plants | water-soluble sugars |
| plant growth and structure | insoluble cellulose, lignin, and other polysaccharides |
| plant decay by microorganisms | cellular proteins, carbohydrates, and growth factors, and release of carbon dioxide and methane |
| sediments | polysaccharide conversion to hydrocarbons under intense pressure and long time periods; methane and natural gas |
| combustion of fossil fuels (human activity) | carbon dioxide gas and volatile organic compounds |
| animal consumption of plants | cellular proteins, carbohydrates, fats, and growth factors; respiration releases carbon dioxide gas |

hazardous substances interfere with a cycle, the food webs that contribute to the cycle also change and ecosystem imbalance may occur. Inevitably an entire community in the environment behaves differently than it naturally would.

Perhaps the most dramatic effect of waste on a biogeochemical cycle occurs when waste pollutes the nitrogen cycle. Chemist Daniel Rutherford discovered nitrogen gas in 1772 and noted that it could not support life in laboratory experiments. The fact that nitrogen by itself does not support life seems surprising since nitrogen is abundant in the body, and makes up 78 percent of the atmosphere. Nitrogen occurs in thousands of compounds and every form of life has nitrogen-containing

## Nitrogen Cycle and Contamination

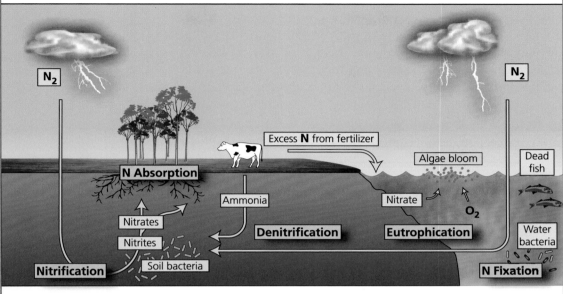

© Infobase Publishing

The Earth's nitrogen cycle usually becomes imbalanced from too much nitrogen entering the cycle rather than too little. A large influx of nitrogen-containing organic matter causes microbial blooms in water and soil. In water, microbial blooms can cause the death of aquatic species. In soil, excess nitrogen interferes with normal nitrogen reactions and plant growth.

compounds. All amino acids contain nitrogen, therefore every protein and every enzyme contain it. The nucleic acids deoxyribonucleic acid (DNA) and ribonucleic acid (RNA) also consist of nitrogen-containing units. The element nitrogen is therefore essential as part of many different types of organic, or carbon-containing, compounds that run cellular systems.

Nitrogen gas normally moves from the atmosphere to the earth through the action of bacteria in soil or water. These bacteria capture gaseous nitrogen in a process called *nitrogen fixation* and incorporate it into their amino acids. The cells then build proteins and other compounds. From there, microbial nitrogen moves upward in *food chains* through plants and animals. When living matter dies and decays, the nitrogen follows either of two paths: (1) nitrogen compounds return to the soil with decayed matter, or (2) specialized bacteria convert nitrogen to gas in a process called denitrification. Wastes from agriculture, industry, and cities all have the potential to interfere with several specific points within each nutrient cycle.

Two different examples of interference with nutrient cycling are provided by combustion engines and agricultural wastes. When combustion engines burn gasoline, they put nitric oxide (NO) into the air, which converts into nitrogen dioxide gas ($NO_2$) and nitric acid ($HNO_3$). Nitric acid is one of the components of acid rain, which damages ecosystems by lowering pH in water and slowing plant growth. Agriculture, by contrast, produces large amounts of animal wastes and fertilizers that both contain a high concentration of nitrogen compounds. When rain washes these compounds into lakes or to the coast, algae in the water enjoy the bounty of new nitrogen and burst into a period of rapid growth called a *bloom*. The algae quickly use up all the other nutrients in the water and begin to die. Bacteria in the water then take their turn at the dining table by feasting on the algal cells. The bacterial growth is so rapid that the bloom uses up all the oxygen in the water. The entire process leading to oxygen depletion is called *eutrophication*. Invertebrates and small animal life suffocate in eutrophied waters, which affects fish that depend on these species as their food source. Next, fishing-eating mammals lose their main source of nutrients and an entire animal community suffers.

Blooms of algae are an obvious sign of an ecosystem out of balance and therefore a sign of pollution, but some forms of pollution are not as easy to spot. One example is the effect of toxic metals in soil. Soil microbes, such as bacteria and fungi, play a critical part in decomposing organic matter, but metal-laced soils kill large numbers of these microbes. As a consequence, a biogeochemical cycle that depends on reactions in the soil may stall. Waste management therefore can have an impact on ecosystems far away and, very often, unseen.

## WASTE MANAGEMENT

Waste management probably began in the Neolithic Age 5,000 years ago when tribes dug drainage channels for carrying waste to the nearest body of water. Water-flushed toilets date back as far as 2500 to 1500 B.C.E. in present-day Pakistan. These conveniences merely removed the sewage to a nearby ditch, but at least people were willing to invent something to carry wastes out of sight. This practice of using natural waters as a disposal mechanism lasted for centuries. The Romans in the sixth or seventh century B.C.E. built the Cloaca Maxima, a great canal that carried sewage from Rome's environs to the sea. In about 9 C.E., the Roman historian Livy described the system's construction by the empire's working classes: "... they felt it less of a hardship to build the temples of the gods with their own hands, than they did afterwards when they were transferred to other tasks less imposing, but involving greater toil—the construction of the bench in the Circus and that of the Cloaca Maxima, a subterranean tunnel to receive all the sewage of the City. The magnificence of these two works could hardly be equaled by anything in the present day." The Roman Empire of 50 to 500 C.E. further improved the water conveyances and sewers; pieces of these structures remain today.

During the Middle Ages in Europe, waste streams happened wherever a person opened a door and threw out their garbage. In 11th-century London the stench of waste forced the development of a new technology in waste treatment, the cesspit. These receptacles were built into the ground near houses and received a daily deposit of household garbage and human waste. Builders intended to make the cesspits leakproof, but sadly they did leak, leading to contaminated waters, orchards, and vegetable and herb gardens. Burying helped dispose of wastes a bit, but buried wastes leaked into underground water that supplied wells. Today's waste management struggles with almost the same set of problems, that is, waste streams are still threatened by accidents and leaks. Waste managers work to prevent the unintended pollution of clean water and soil with waste, and they also continue to find better ways for removing waste from people's lives. The historian Jon C. Schladweiler on his *History of Sanitation* Web site (URL: http://www.sewerhistory.org/) described the progress of waste management for the last 150 years: "In 1847–48, the British Parliament adopted a sanitary code that applied to all of England and Wales—but not London. The sewer commissioners heard about the attributes of the sewerage systems developed by their ancestors on the Isle of Crete in Greece; those systems served as examples for the designers of the new sewers soon to come in the London area."

The main goal of waste management has not changed in the last centuries: It strives to manage waste streams. After waste streams have been properly managed to prevent leaks into the environment, and thus pollution, new technologies have emerged for treating the waste and making it less hazardous to ecosystem health.

Modern waste treatment consists of physical, chemical, or biological methods. Physical treatment captures materials so that they cannot migrate and pollute uncontaminated places. *Encapsulation, filtration,* settling by gravity, adsorption, and stabilization are some examples of physical treatment. Chemical waste treatment methods convert hazardous compounds to harmless compounds through chemical reactions. Often chemical treatment is done right at the site where contamination occurs and so it may be referred to as cleanup/treatment. Examples of chemical cleanup/treatments used today are chemical oxidation and thermal treatment. Thermal methods treat chemicals by destroying them in intense heat. Incineration and vitrification represent two types of thermal waste treatment: (1) incineration reduces waste to ash; (2) vitrification converts waste to a stable glasslike form. Biological treatments use microbes or plant life to degrade wastes or at least hold them in place so they do not move in the environment.

Each waste treatment method must offer cost advantages yet not injure the environment. For these reasons, waste management professionals must understand new technologies for choosing the best method for a particular task. Each waste management choice also involves special aspects such as waste transport, recycling, chemistry, biological restoration of land, and environmental law. All of these specialties play a role along the course of a waste stream until the wastes reach their final destination. Proper waste transport consists of the delivery of waste loads from their source to a final treatment or disposal site, and of all the different aspects of waste streams, transport has a visible impact on community confidence.

## WASTE TRANSPORT

People have always desired efficient and fast waste removal methods even if they had not yet mastered the technology to provide this benefit. Without a functioning waste removal system, societies confronted the hazards of infectious materials entering their communities. These hazardous materials came from animals, other people, and physicians' treatment of the sick and dying. Transporting *infectious waste* away from a healthy population helped stop the spread of disease. Meanwhile, people disposed of nonhazardous and noninfectious wastes by the most convenient method at hand. Today, hazardous and nonhazardous waste transport has become more sophisticated and efficient. Yet the basic concept remains the same as it was centuries ago: Remove the materials as quickly as possible from people to lessen potential health hazards.

Nonhazardous waste transport is done by companies serving a single town or a certain region. The customer (the town or region) sets up a contract with a local waste hauling company to manage its solid waste stream, its MSW. Across the United States, waste haulers daily devote almost 500,000 vehicles to pick up and transport MSW. Thirty years ago

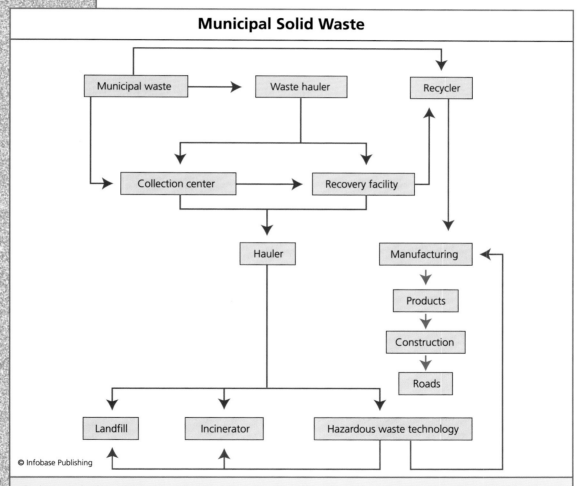

**Municipal Solid Waste**

© Infobase Publishing

The best methods in MSW management take advantage of new technologies in waste type separation, recycling, reclamation, and disposal. Sustainable MSW management strives to find innovative ways of rerouting as much waste material as possible for new uses and to minimize the total amount to be landfilled or incinerated.

these waste haulers threw nearly any type of household, office, or restaurant waste into their trucks. Today waste hauling companies work with communities to manage their waste streams. This process usually begins by separating reusable (recyclable) from nonreusable items and keeping hazardous substances apart from the general MSW.

Waste haulers take each daily load to a centralized site called a *treatment, storage, and disposal facility* (TSDF). TSDFs are licensed facilities that are responsible for managing a community's solid waste streams. TSDFs recover reusable items that have not already been sent to recycling. They also remove any hazardous materials that accidentally became part of the waste stream. Depending on the town and its separation/recycling needs, a TSDF may take in the additional following items: used packaging, bottles, newspapers, furniture, clothing, appliances, and yard trimmings, along with household trash and restaurant garbage, much of it food scraps. Some towns separate out many of these items before the waste hauler picks them up, but in other towns, the TSDF must separate the components of MSW. The main components in today's MSW handled by TSDFs are listed in the following table.

| THE EPA's ESTIMATE OF MSW BEFORE RECYCLING (Percentage of the 250 Million Tons [227 million metric tons] Produced per Year) | |
| --- | --- |
| **WASTE COMPONENT** | **PERCENT** |
| paper and paperboard | 34 |
| yard trimmings | 13 |
| plastics | 12 |
| food scraps | 12 |
| metals | 8 |
| rubber, leather, textiles | 7 |
| wood | 6 |
| glass | 5 |
| other | 3 |

In most communities, construction debris, nonhazardous industrial wastes, and wastewaters are handled and transported separately from MSW. Nonhazardous materials usually go to landfills, and wastewaters flow to specialized treatment plants, which are described in chapter 7. Once the TSDF has removed recyclables and hazards, it consolidates the rest of the waste into larger loads to go to an incinerator or a landfill.

The EPA has instituted a program called WasteWise in which the agency works in cooperation with the waste management industry to streamline waste transport. Within WasteWise guidelines, haulers focus not on the total MSW they transport but on methods for reducing the amount they transport. The EPA's WasteWise Web site states the ultimate benefit of this approach: "Waste reduction makes good business sense because it can save your organization money through reduced purchasing and waste disposal costs." Local governments, schools, and businesses also follow the tips published by WasteWise to lower the costs they pay to waste haulers. Will waste haulers accept a program designed to reduce their profits? The EPA provides online resources that explain better methods for sorting and recycling, while ensuring waste management companies' profits do not decrease. Eventually waste management's primary focus will change from the tons of MSW transported each week to the innovations that reduce waste.

Even with a new viewpoint in waste transport, change comes slowly. Community waste management companies have adapted well to the ideas put forth in the WasteWise program, but industry lags behind. Industrial waste management remains chiefly an issue of transport and not reduction. These industrial waste loads amount to hundreds of tons that move mostly by truck on common thoroughfares shared with other drivers. Railroads handle no more than 20 percent of the load and a small amount also moves by ship. Due to the enormous tonnage of industrial wastes crisscrossing the nation each day, transport remains one of the waste industry's biggest challenges.

When industrialization expanded in the 1930s and grew until the early 1980s, waste haulers had little incentive for thinking of efficiency. They carried away any and all material for a fee, whether the substances were hazardous or not. Profits accrued based on the total volume they transported or the number of pickups they made, and speed rather than careful handling equaled profits. Road accidents, spills, and improper disposal, plus unlawful dumping, became more and more frequent. Congress responded to the growing problem of improper hazardous waste transport with the *Resource Conservation and Recovery Act* (RCRA). The act established a new philosophy for hazardous waste: cautious handling, transport, and disposal with consideration to the environment at every step along the way. To do this the RCRA mandated that all hazardous waste be tracked from point of origin to its final disposal site.

The RCRA's purpose was to increase the safety of carrying hazardous wastes through neighborhoods. According to the new law, only licensed haulers could transport the large hazardous waste loads produced by industries. Unfortunately, hazardous waste transport soon became more, not less, dangerous. Instead of paying a licensed waste hauler to comply with the regulations, companies began illegal dumping. In a 1980 *Time* magazine article, the correspondent Ed Magnuson wrote, "One day a field in Illinois was empty; a week or so later, it contained 20,000 barrels of dumped wastes." Magnuson described but one of hundreds of such instances. Companies used their own trucks to unload wastes in rural areas during the night. Large tanker trucks were fitted with valves for secretly releasing liquids onto the road as they traveled. Sometimes drivers simply took waste-filled trucks outside of town and abandoned them. As noxious chemicals accumulated in the environment, Congress added amendments to the RCRA to further control hazardous waste transport. Today, the EPA and the U.S. Department of Transportation (DOT) share responsibility for overseeing hazardous waste transport, and the RCRA gives them the authority to enforce and fine lawbreakers.

The DOT classifies tank trucks and rail tank cars based on the type of waste they carry: combustible and flammable liquids with low vapor pressure (fuel, gasoline), flammable liquids with high vapor pressure (toluene), corrosives (acids), liquefied compressed gases (chlorine, propane), or refrigerated compressed gases (oxygen). A USDOT number must appear on the tank, indicating the type of waste inside. For example DOT-412 describes a corrosive material such as hydrochloric acid. The tanks themselves are designed to withstand corrosion from within and to prevent waste materials from igniting, exploding, or reacting with air or moisture. Tanks are usually made of steel or aluminum alloy, and newer designs might include stainless steel, titanium, or nickel. The type of tank, indicated by a motor carrier (MC) number, is displayed on the vehicle, and it must correspond to the USDOT number. This assures that the correct vehicle carries the hazardous waste it is designed to carry.

An important aspect of waste management relates to waste transport. This barge carries tons of waste down the Mississippi River. *(WQPT)*

In the United States, the National Hazardous Materials Route Registry (the Registry) designates roads that hazardous material (*hazmat*) carriers may use for transporting waste. Some hazmats use only roads consisting of certain construction specifications and level of maintenance. The Registry periodically inspects these roads and updates the list. It drops poorly maintained roads from the list and adds new and safer roads. Some of the criteria the Registry uses in evaluating roadways are: highway construction, population density nearby, terrain, availability of emergency response teams, local weather, local environmental factors (earthquakes, flooding, high winds), and accident statistics.

Radioactive waste receives special attention from the DOT whether it is moved by truck, rail, or ship. Government agencies are responsible for the details of each shipment of radioactive wastes according to DOT-enforced laws. The responsibilities of the agencies are as follows:

- truck transport—the Federal Motor Carrier Safety Administration
- rail transport—the Federal Railroad Administration
- ship transport—the U.S. Coast Guard

The U.S. Department of Energy (DOE) designates radioactive waste shipped across the nation in the following three categories: high-level, low-level, or transuranic. High-level wastes contain the most radioactive materials produced by the nuclear industry; low-level wastes consist of lower activity materials in large loads of nonradioactive matter; and *transuranic wastes* consist of the by-products of nuclear substance manufacture. Once the wastes have been put into these categories, each transporter follows DOT rules pertaining to the type of material to be carried. In addition to federal laws, large shipments (several tons) of high-level materials must follow state and local regulations. These regulations apply to all the areas through which a shipment travels.

Even with added regulations on radioactive waste transport, communities near roads and railroads that receive these transports have been concerned about potential accidents. In 2003, a team of environmental organizations filed a lawsuit to halt radioactive shipments in California and Ohio, and Charles Weems of Washington Physicians for Social Responsibility said in an issue of *Waste News,* "Public health, especially the health of children, is placed at risk by trucking radioactive waste shipments that expose people to unnecessary radiation." The DOE has forecasted that 10 to 15 shipments of high-level waste per year will travel across U.S. state lines for the next several years. Furthermore, the DOE predicts that these shipments will increase to about 300 annually by 2010 and up to 1,700 by 2015. The agency has tried to assure the public that the process will be safe. The DOE has stated, "The department must ship waste according to strict federal regulations. The waste will be transported in heavily shielded casks certified by the Nuclear Regulatory Commission (NRC) along approved transportation routes." Transuranic waste, which has lower radioactivity than high-level wastes, will also require thousands of shipments per year through 2015. Trucks and railroads will probably remain the main modes of transport for all of these shipments.

# LANDFILLING

Dumping nonradioactive waste in remote areas outside of populated areas has been the cheapest and easiest answer to disposal for thousands of years. Garbage dumps have been discovered near many of the oldest sites studied by archaeologists; items found there have helped them learn details of early tools and implements. *Landfilling* does not treat waste,

Landfills have been a main waste disposal method for centuries. Landfills are now reaching full capacity in many places, so waste managers now use landfills to complement waste treatment rather than serve as a sole solution for waste disposal. *(Envirowise)*

however, it simply serves as a long-term storage for it. Disposal in landfills has recently begun to decrease each year because on-site cleanup/treatment has improved. Despite this decrease, these sites still serve as convenient disposal for certain nonreusable, nonrecyclable items and avoids the potential hazards of transport.

When people awakened in the 1970s to the decline of the environment, concerned scientists took a close look at landfills. They found sites filled with unidentified mixtures of hazardous materials and chemicals leaching into nearby soils and *groundwaters*. Additional wastes washed from stockpiles with each rainfall and made their way to waterways and estuaries. A 1976 *New York Times* article reported on one of many examples when it noted, "The state has ordered New Jersey's only chemical landfill to close within ten days on charges of continued violations of environmental standards." New environmental laws began to address the hazard of poorly managed landfills. Today landfills belong to classifications according to the type of waste they accept, as shown in the following table.

| TYPES OF LANDFILLS IN THE UNITED STATES | |
|---|---|
| LANDFILL TYPE | MATERIALS ACCEPTED |
| cleanfill | clean excavated soil and inert (nonreactive) materials (wood, metal, glass, paper, etc.) |
| industrial waste | nonhazardous wastes from local industries |
| industrial-municipal mixed | industrial nonhazardous wastes and MSW |
| municipal solid waste (sanitary landfills) | MSW and other inert materials |
| hazardous waste | substances designated by the EPA as hazardous |

MSW landfills are the most common type in the United States, accepting about 55 percent of all MSW. (About 30 percent of MSW is recycled or composted and 15 percent is incinerated. *Compost* is a mixture of organic matter allowed to decompose over time.) Because so many landfills were built in the United States prior to the 1980s—there were 8,000 by the end of the decade—many municipalities have little land left for additional sites, and they build fewer landfills today. Many of the existing landfills have been covered over and closed according to regulations set by the RCRA. Despite the closures, more than 2,000 landfills remain in operation, and these sites receive more careful oversight than landfills received in the 1970s–80s. Landfill operators now employ new techniques in landscaping the waste site and also use improved containment methods to prevent materials from leaching into adjacent land and water.

Modern landfills designed for MSW are called sanitary landfills because of their leak prevention systems that keep the surroundings clean. Most sanitary landfills contain a multilayered underlining of compacted soil and leak-proof sheets of plastic. In the past, linings consisted of dense clay one-foot (0.3 m) thick plus plastic sheets, but the clay often cracked and chemicals escaped. Modern sanitary landfills use synthetic liners made of *high-density polyethylene* (HDPE) plastic with a thickness of at least 1.2 inches (3 cm). Larger sheets minimize the number of places where sheets must be joined together and special welding provides leak-proof

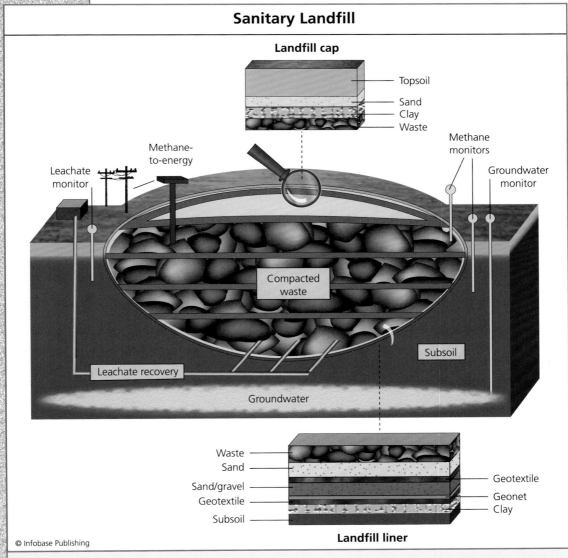

**Sanitary Landfill**

**Landfill cap**

Topsoil
Sand
Clay
Waste

Methane-to-energy

Methane monitors

Leachate monitor

Groundwater monitor

Compacted waste

Subsoil

Leachate recovery

Groundwater

Waste
Sand
Sand/gravel
Geotextile
Subsoil

Geotextile
Geonet
Clay

**Landfill liner**

© Infobase Publishing

Contemporary sanitary landfills use materials that hold in hazardous leachates and gases, but also help regulate gas buildup and temperatures inside the compacted waste. New landfills also incorporate sensitive monitoring systems to detect leaks, and some landfills have equipment to capture methane gas for use as an energy source.

connections between the joints. HDPE is also stronger than the plastics used in the past, but some chemicals degrade HDPE, so it does not provide fail-safe protection. Even the best constructed landfills require close monitoring to assure that their contents stay in place.

Advanced landfills depend on composite systems that consist of containment layers interspersed with monitoring devices. Composite systems also contain drainpipes placed between liners to draw *leachates* from the load. The pumped leachates then receive treatment to remove any hazardous chemicals. Landfill operators monitor sensors to detect leachates entering the soil or groundwaters and check for excess, ignitable methane gas produced by microbes. The best sanitary landfills include a cap on top of the waste load to prevent rain from entering the load and so protect against runoff or high winds. Innovative cap arrangements consist of soil layers alternating with synthetic filters designed to control the release of gases.

Containment liners for hazardous wastes include sand and gravel layers alternated with plastic netlike liners called geonets. Flexible *geomembranes,* made of the plastics polyvinyl chloride or HDPE, or fabrics may also be part of the structure. Geotextiles (specialized fabrics) also help by trapping small particles to prevent clogging while allowing water to filter through. All of these innovations resist breakdown by chemicals and damage from repeated freezing and heat.

Ordinary household and restaurant garbage decomposes within landfills in stages. First, aerobic (oxygen-requiring) bacteria and fungi digest degradable matter. In the process, they consume oxygen and produce carbon dioxide, water, and other by-products of their metabolism. The decomposition process also produces heat (122–158°F [50–70°C]) inside the waste load. After about two weeks, the second stage begins in which the oxygen is gone and anaerobic bacteria predominate. These microbes cannot live in the presence of oxygen and their unique metabolism produces more carbon dioxide, methane, and organic end products. Carbon dioxide and methane make up more than 90 percent of the gaseous compounds released from landfills, and they contribute to the atmosphere's greenhouse gases. Anaerobic end products also emit unpleasant, though harmless, odors that tend to annoy communities living near even the best-managed landfills.

Landfill methane can serve as an important energy source in a process called *waste-to-energy* (WTE). About 425 WTE landfills operate in

43 states with at least as many additional ones planned for the near future. Energy produced this way from landfills averages 0.8 megawatts for each ton of MSW, and since 2003 landfill methane has been traded on the Chicago Climate Exchange and the European Climate Exchange. The Chicago Exchange allows North American corporations or towns that reduce their emissions below a set limit to sell emission credits to other organizations, or save them for the future. Meanwhile, companies having a difficult time meeting emission limits purchase credits through the exchange. Dave Miller of the Iowa Farm Bureau Federation discussed methane credits in a 2006 *Brownfield News* article, saying, "With natural gas prices where they are, the energy system will pay for itself."

Many landfills now ban specific wastes as an extra safety measure against dangerous leachates or potential reactions within the waste load.

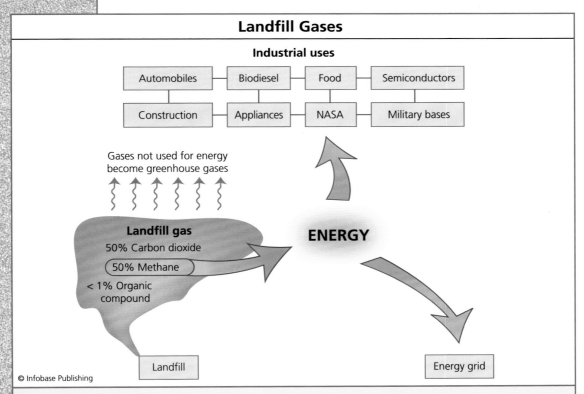

# Landfill Gases

### Industrial uses

| | | | |
|---|---|---|---|
| Automobiles | Biodiesel | Food | Semiconductors |
| Construction | Appliances | NASA | Military bases |

Gases not used for energy become greenhouse gases

**Landfill gas**
50% Carbon dioxide
50% Methane
< 1% Organic compound

**ENERGY**

Landfill

Energy grid

© Infobase Publishing

Biochemical reactions inside landfills emit greenhouse gases, but new technologies now capture one of these gases, methane, and reroute it as an energy source for a variety of industries.

The following banned items must follow a different waste stream to final disposal: electronics, mercury-containing items, batteries, fluorescent bulbs, and partially filled aerosol cans. Electronics may contain *heavy metals* or toxic flame retardants. For instance, mercury works in electrical switches, thermometers, barometers, and some medical devices. Aerosol cans contain hydrocarbon propellants such as the greenhouse gas carbon dioxide, propane, or butane, and these cans often contain paints and solvents dangerous to the environment. In summary, landfill management requires some knowledge of chemistry and the components that make up the enormous diversity of waste items. The following sidebar, "Case Study: The Birth of a Throwaway Society," discusses the reasons why waste seems to build faster than people can treat it.

## SEPARATION AND TREATMENT TECHNOLOGY

Recovery of reusable materials relies on TSDFs to separate reusable waste from nonreusable materials. TSDFs divide materials based on chemical

### CASE STUDY: THE BIRTH OF A THROWAWAY SOCIETY

The United States produces enough wastepaper each year to build an 11-foot (3.3 m) wall coast to coast. People throw away 2 billion pieces of junk mail each year. Electronics that are barely a year old end up in landfills. Convenience foods generate tons of packaging. These are hallmarks of consumerism in Western culture. So much consumption-to-waste takes place in Western, industrialized economies that the term *affluenza* has been proposed to describe this unsustainable addiction to consumption, usually overconsumption. The United States and Canada may suffer more than any other nation from this affliction; the United States has less than 5 percent of the world's population yet produces one-third of its solid waste. Overconsumption is devouring resources and producing a growing mountain of waste.

Sociologists propose several reasons why consumer product waste has grown tenfold in the last 100 years. First, wealth allows people to feel less inclined to reign in consumption to save money. Second, busy schedules lead to the use of convenience products, which generate large amounts of waste. Americans discard 130 million cell phones and 50 million computers each year. Third, innovations, especially in electronic products, make existing products become

composition and potential health hazards and then further group materials by treatment method or reuse potential. These facilities often also have the capability to chemically neutralize some hazardous substances. Overall, TSDFs play a critical role in separating solids from liquids, oils from aqueous fluids, incompatible materials from each other, and materials requiring special treatment. Industrial wastes represent a complex collection of materials that TSDFs must understand in order to send them to the correct type of treatment facility. The main categories of industrial wastes managed by TSDFs are shown in the table on page 30.

Local TSDFs provide each community with information on waste sorting, what constitutes a hazardous waste, and where waste should be taken for disposal. TSDFs often provide information on how to sort carpeting, boxes, clothing, furniture, appliances, window glass, and other materials before pickup. At the facility, TSDF workers conduct additional sorting and transfer wastes into secure containers. Hazardous liquids fill 55-gallon (208 l) drums or larger mobile tanks made of steel, plastic, or fiberglass. Some TSDFs maintain open storage piles, lined and monitored similar to landfills. Once hazards have been stored in a safe container, the

obsolete faster than ever before. The London *Times* reporter Richard Morrison wrote in 2007, "Does anybody [today] buy a car, a washing machine, even a toaster, in the expectation that it will last a decade? As for computers, mobile phones, iPods and all the other electronic paraphernalia of our gizmo-fixated age, well, the philosophy among manufacturers seems to be that since [people] will surely want to 'upgrade' every twelve months, there's no reason, let alone obligation, to make products that last any longer." Finally, new electronics, furnishings, and fashions may meet an emotional need in a consumer-based society, which only contributes to how quickly things go obsolete.

Reversing this trend will be difficult. Success will come from a combination of personal choices in purchasing and innovations from industry that either reduce waste or make waste more recyclable. Community recycling programs have made strides in reducing paper, glass, and metal wastes. Industries that make fabrics, clothing, furniture, and construction materials must offer products designed for similar sustainability. Unfortunately, making conservation a priority over convenience and low cost may prove to be a difficult step for individuals and businesses alike.

| TYPES OF INDUSTRIAL WASTES | |
|---|---|
| **INDUSTRY** | **TYPES OF WASTE IT PRODUCES** |
| medical | medical nonhazardous, radioactive, infectious |
| services | paper, electronics, food, furniture, clothing, packaging |
| education | paper, electronics, furniture, chemicals |
| utilities | cooling waters, metals, disinfectants, electronics |
| manufacturing | packaging, raw materials, process water, leftover products |
| construction | wood and other building materials, wire, insulation, paints |
| transportation | fuel, tires, vehicles, food, seating, furniture, electronics |
| extraction | mine and mill tailings, acids, equipment, metals |

TSDF arranges transport of each type of waste to an appropriate recycling, disposal, or treatment site. In these ways, each TSDF serves as a central point in waste stream management.

Industrial wastes require more work in sorting and separating than MSW, but some communities have begun to master the difficult task of managing and recycling materials left over from industrial processes. The following sidebar, "Case Study: Denmark—A Model in Waste Management," describes innovations in sustainable uses of industrial wastes.

## THE SALVAGE INDUSTRY

The salvage industry recovers solid wastes from manufacturing firms and sells the wastes as raw materials to other businesses. Salvaging may be considered the first true recycling operation before the terms *reuse* and *recycling* became popular.

Scrap timber and metals have been recovered for reuse since 2000 B.C.E. Prior to the Industrial Revolution, every society depended on a

## CASE STUDY: DENMARK—A MODEL IN WASTE MANAGEMENT

Denmark's coastal city of Kalundborg has developed one of the world's premier waste management and energy-sharing plans. The plan took root after a severe water shortage in the 1980s–90s, which stressed local industries. The term *industrial symbiosis* was coined in Kalundborg for a process whereby city and industry activities were blended into a resource-sharing system. Kalundborg works much like food webs in nature, explaining why it has been called an industrial ecosystem.

A coal-burning power plant serves as Kalundborg's central point. The town, its manufacturing, and nearby agriculture connect to each other through this energy-generating center. Wastes from several of the enterprises serve as raw materials for others within the system. For example, a desulphurization operation within the oil refinery converts sulfur into ammonium thiosulfate fertilizer for local farms. Meanwhile, a cement manufacturer uses the power plant's excess ash. Energy producers transfer excess heat to energy consumers, such as the municipality, and the entire system recycles energy at the same time it conserves raw materials. The head of the U.S. company Triad Energy Resources, Inc., Michael Daley, observed in the *New York Times* in 1999, "If companies were smart, they'd all locate near sources of waste."

Kalundborg has reduced its water consumption by 25 percent by recirculating water among the partner companies; annual water savings reach 71 million cubic feet (2 million m³) of groundwater and 35 million cubic feet (1 million m³) of surface water. Excess steam circulates through the system to enable each company to reduce oil use by substituting part of its energy needs with steam.

Twenty similar industrial ecosystems now operate in various parts of the world using Kalundborg's model. Kalundborg's power plant continues innovating by experimenting with *biomass* and wind energy in preparation for its eventual conversion from coal to renewable energy sources.

Europe now leads the world in the amount of industrial wastes it sends to some form of waste-raw material exchange similar to Denmark's.

*(continues)*

*(continued)*

At least one-third of all wastes in Europe go into these systems. The United States, by contrast, lags behind in the world of industrial ecosystems; only about one-tenth of industrial wastes in the United States go to Kalundborg-like exchange systems.

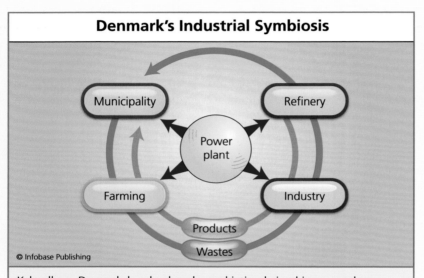

## Denmark's Industrial Symbiosis

© Infobase Publishing

Kalundborg, Denmark, has developed a symbiotic relationship among the municipality, the waste treatment plant, the water treatment facility, and a centralized power plant. Energy in the form of steam circulates through several industries as does treated water to assure that energy and water waste have been minimized. Some industrial wastes serve as raw material for other industries and all unusable waste then goes to the municipal treatment plant for final disposal.

cottage industry of those willing to comb through rubbish to find materials of value. In the 1500s, scrap dealers recovered iron from copper mining sites; in 1588, Queen Elizabeth I decreed the collection of discarded rags for use in papermaking. By the late 1800s, British workers sorted waste by hand and made a living recovering and selling any reusable materials they found. Salvaging grew in importance when the Industrial Revolution began in the 18th century and made mechanized manufacture a standard way of doing business. Machines made products significantly faster than manual labor could and, as a result, employees no longer had time to save extra materials and put them back into the production line. Increased speed in manufacturing created an inevitable and mounting pile of industrial scraps. Entrepreneurs soon descended on the industrial wastes as waste sifters had a century earlier.

Automotive parts and construction wastes are today's most lucrative areas in the salvaging industry. Automotive salvaging recovers car, truck, and tractor parts for resale or for rebuilt parts. Scrap cars, scrap parts, mercury light switches, metals, glass, and other materials find their way into millions of new or rebuilt vehicles. For instance the steel industry recycles 14 million tons (12.7 million metric tons) of steel from vehicles, an amount equivalent to about 13.5 million new cars. Recycling businesses make the shiny model sitting in an auto showroom one of the world's most recycled consumer products.

Japanese, German, and U.S. automakers in 2006 collaborated to create the End of Life Vehicle Solutions Corporation (ELVS). This collective effort further reduces the waste generated by the automotive industry through more efficient recycling plus innovations for reusing scrap metals. One such device is a shredder that can turn an entire automobile body into small pieces in a matter of minutes. Skip Anthony, the sales manager for the American Pulverizer Company, told *Recycling Today* magazine in 2007, "We listened to the mid-sized scrap dealers express a need to shred and developed our super heavy-duty 60-inch machines to fill this need."

At present, the ELVS has turned its attention to the recovery of mercury from switches, which the EPA requires must be removed before salvagers crush a vehicle. The ELVS 2006 Annual Report summarized its progress in switch recycling: "The first step in implementing an ELVS switch collection program in a state consists of developing a list of scrap recycling facilities, vehicle recyclers, salvage yards, and auto shredders to participate in the program. ELVS sends collection buckets with educational, training, and program materials to those on the list. Participants remove switches from end-of-life vehicles and put them in the bucket. When the bucket is full . . . participants ship the container free of charge to the Environmental Quality Company (ELVS's waste handler)." The EPA provides further guidance through the National Vehicle Mercury Switch Recovery Program, which, despite the mercury recovery program, estimates that 67 million mercury switches are still in use in older model cars.

Today's salvage industry has become a sophisticated dealer in valuable materials. Scrap metal is one of its most important markets, led by aluminum, zinc, magnesium, lead, nickel, stainless steel, and copper and brass, known as the red metals. Salvagers also recover iron from cast-iron products, railroad tracks, and the steel inside tires. Other salvagers specialize in glass, paper, and plastics.

The U.S. salvaging business has grown into a sophisticated arm of the recycling industry. Salvagers recover metals and melt them, as the molten bronze shows here, and then send the purified ingots to the metal industry. Mercury, palladium, platinum, brass, copper, and nickel provide examples of additional salvaged metals. *(Art and Perception)*

Outside the metals industry, salvagers target construction waste and demolition scraps for sale to builders of new houses and for remodeling. Many builders and architects are especially interested in items from very old houses because they supply a niche market seeking early 20th-century fashions: Old light fixtures, glass doorknobs, faucets, mantels, and ironwork are valuable commodities. In 2007, salvager Steve Drobinsky explained in *This Old House* television show, "This week, one of the oldest mansions in San Francisco was being remodeled and they were removing marble sinks, cast-iron fire screens with stags and forests, a hand-carved walnut mantel—one leaf over two feet long, all hand carved! I mean, what could you get that could be better than that?"

## CONCLUSION

The amount of waste generated in the world is growing. It is greatest in Western cultures, and sociologists associate waste volume to the level of affluence in a society. Managing waste is one of the first steps in ensuring an ecosystem functions properly because metabolism is affected by the buildup of waste products.

Wastes are classified as hazardous or nonhazardous. Hazardous wastes are a concern because they have the potential to damage plant or animal health. The hazardous materials may be further classified in a number of ways: by chemical composition, by source, by the industrial activity that produced it, or whether it is biological or chemical in nature. Preconsumer wastes are those produced during the manufacturing process, and postconsumer wastes consist of extra, unused products plus discarded packaging. Waste managers assess all these many types of waste to determine how the materials are to be treated, transported, or disposed of.

Waste moves from its source to its final disposal site in a path known as a waste stream. Any waste stream can fall victim to accidents that cause spills into the environment. Hazardous and nonhazardous materials escaping their normal waste streams damage ecosystem health and can impose immediate health hazards on humans, animals, or vegetation. Therefore, waste transport is a critical aspect of maintaining waste streams. In the United States, trucks carry most of the nation's wastes to treatment and disposal sites, and any waste transport vehicle—truck, rail, or ship—must abide by strict government safety regulations.

Landfills are an alternative to waste treatment. Cleanup/treatment methods are becoming more efficient and, as a consequence, the number of landfills is decreasing in the United States. Landfills still serve a purpose in accepting wastes that cannot be treated. Modern landfills contain advanced liners and caps, which have greatly reduced leaching and erosion.

Two different industries participate in making waste streams safer and more efficient. The first is the TSDF, which cooperates with communities and waste haulers to sort wastes. In doing this, the TSDF reduces the volume that must be treated or landfilled and increases the amount of wastes that can be recycled. The second industry is salvaging, which reduces total waste volume by recovering specific components and putting them back into other industries as raw materials.

Waste management has grown into a vital aspect of environmental protection. Waste managers today have responsibilities beyond picking up and hauling garbage. They must manage and maintain safe waste streams. Additionally, waste management includes several specialty areas dealing with wastes and their effect on the environment. These aspects can be divided into three general areas: (1) proper waste handling, (2) waste reduction, and (3) waste recycling. In summary, waste management has become a crucial part of achieving natural resource conservation.

# ELECTRONIC PRODUCTS AND METALS

Consumers and businesses discard 50 million tons (45 million metric tons) of electronic products annually around the world. These electric- or battery-powered items make up 5 percent of global municipal solid waste (MSW), and, in the United States, a disheartening 85 percent of waste electronic products end up in landfills.

Electronic waste, or e-waste, is any broken or unwanted electronic device or electrical appliance. E-waste is also sometimes referred to as WEEE, or waste from electrical and electronic equipment. Examples of e-waste are: televisions, computers, monitors, servers, printers, scanners, radios, CD and DVD players, walkie-talkies, calculators, digital cameras, stereo equipment, video games, global positioning devices, microwave ovens, large and small kitchen appliances, and telecommunications devices such as telephones, answering machines, cell phones, facsimile (fax) machines, personal digital devices, and portable music players. Batteries other than household alkaline varieties also belong to e-waste.

Innovations take place in the electronics industry at a dizzying speed, making e-waste one of the fastest growing waste categories in the world. Reporter Kent Garber wrote in a 2007 issue of *U.S. News and World Report,* "The environment, in turn, is suffering the fallout. The dumping of electronic waste is contaminating groundwater, polluting the air, and endangering people in alarming numbers. The Environmental Protection Agency estimates that 2.6 million tons [2.4 million metric tons] of 'e-waste' are produced in the United States each year, or roughly 20 pounds [9 kg]

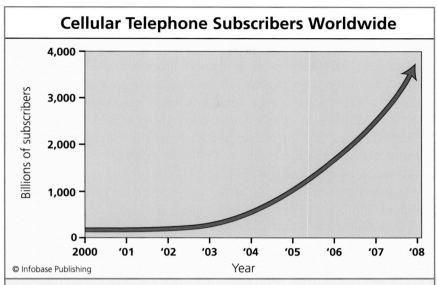

**Cellular Telephone Subscribers Worldwide**

Cell phones contribute to the rapid growth of e-waste. Cell phone users have increased almost exponentially in the past several years with no signs of slowing. In 2010, cell phone ownership worldwide had passed 4.6 billion subscribers and continued to increase.

per person. Much of it is generated during the holidays, when consumers replace outdated units with newer models."

Household appliances, discussed in the sidebar on page 45, "White Goods," contribute the largest single share of e-waste, particularly because these are often large, bulky items. Household appliances also contribute a vast array of hazardous and nonhazardous components to waste streams. Some of the major wastes from household appliances are hazardous and nonhazardous metals, plastics, glass, refrigerants, and oil. Information and telecommunications devices also make up a significant portion of e-waste, mainly from discarded telephones, cell phones, and wireless communication devices. Consumer equipment (televisions, stereos, DVD players, MP3 players, etc.) makes up the remaining portion of e-waste.

E-waste has been associated with the world's wealthiest economies where individuals may own several electronic devices. The U.S. Census Bureau reports that at least 60 percent of U.S. households own at least one computer; libraries, businesses, and universities own additional millions of computers. Though the United States leads the world in computer and other electronics ownership, the worldwide growth of e-waste is shifting.

During an eight-year period up to 2000, computer purchases doubled in the United States, yet during the same period they tripled in Switzerland and increased more than six times in Brazil. In China, computer use increased by 1,052 percent—more than 11 times the country's computer use—within seven years! India's fast-growing economy has had a similar effect on e-waste accumulation. India's increase in purchases of phones, computers, printers, audio equipment, and appliances has resulted in more than 146,000 tons (132,450 metric tons) of e-waste thrown out in that country annually. This amount is expected to triple by 2015.

The telecommunications industry and global data transfer systems have advanced perhaps faster than any other technology in the electronic world. In addition to the United States, this technology has made enormous growth in the following Western economies: Australia, New Zealand, United Kingdom, Canada, Western Europe, and Scandinavia. The Chinese market for telecommunications devices may soon dwarf all others; China is the world's largest cell phone market and has 20 percent of all cell phone users.

In current electronics technology, innovations advance faster than the expected lifespan of products. When older models are thrown away to make room for newer models, a product's useful life becomes shortened. For example, a cell phone's expected lifespan is currently about 14 months even though the device can work well for much longer. For each portable product, used batteries also make up a special portion of e-waste, as discussed in the sidebar on page 42, "Household Batteries."

E-waste complicates waste management because electronic product recycling makes up only a small portion of total waste recycling. Most countries today recycle less than 20 percent of discarded electronics. The reasons for this poor rate of recycling range from lack of information to inconvenience and even laziness. The consumer researcher Stephen Baker stated bluntly in an article published in 2008 by *Reuters* news agency, "People aren't going to do it because people are lazy. When it comes right down to it there are no incentives." In the same article, the Chelmsford, Massachusetts, recycling coordinator, Jennifer Almeida, admitted, "It's a bit of a madhouse," when describing the line of cars queued at a recycling center. "It's not convenient for residents and it's just not Earth friendly. It's a whole lot of cars burning a whole lot of fuel."

This chapter reviews the problems associated with high-volume purchasing and the successes and failures—and challenges—in today's recycling of electronic products. It explains how e-waste harms the environment and discusses types and components of e-waste, and the methods being used for reducing diverse materials contained in e-waste.

## COMPONENTS OF E-WASTE

The components of electronic products create an environmental hazard, but before these components can cause their harm, e-waste fills public landfills because it has never been part of an efficient recycling program. Even worse, tons of e-waste are discarded in illegal dump sites where no monitoring takes place. Older model computers, monitors, and televisions are bulky and take up landfill space, and, as they weather, they begin to leak a steady stream of hazardous materials into the environment.

Electronic components create a troubling dilemma for waste managers because of the sheer variety of materials. Computers, for example, have a durable outer shell, usually made of plastic, and various additional plastics in liquid crystal display screens, supports, switch components,

Obsolete computers and support devices do not lend themselves to efficient and safe recycling. Several states have put forward bills that would force computer sellers to offer pickup of old devices and institute a recycling plan. *(Envirowise)*

and internal joints. Inside each computer, a circuit board contains micro-processors and graphics and memory cards made of plastics and metals, including lead soldering. The inner workings also contain beryllium in the motherboard, cadmium in semiconductors, chromium in discs, lead in the batteries and monitors, and mercury in batteries and lamps. The addition of a hard drive, disc drive, fan, and power source unit make computer dismantling and recycling a very complicated task. Because these tasks are complicated and consist of several steps, electronic recycling is rather inefficient and slow compared with other types of recycling. The task is daunting. Greenpeace, a global environmental watchdog and action organization, estimates that a single computer contains more than 1,000 toxic materials.

Computer recycling businesses find it difficult to keep up with the growth in e-waste because electronic devices enter the marketplace faster than efficient machinery for handling and dismantling e-waste. E-waste streams today rely on significant manual labor, which has its own unique set of energy requirements. Manual labor costs energy in the form of heating or cooling, ventilation systems, water supply and plumbing, lighting, heat for food services, and other support equipment. Often a good deal of the support equipment involves more computers!

Fax machines and printers are examples of computer peripheral devices that contain environmentally hazardous inks; ink cartridges now make up a specialized area within e-waste recovery. To achieve a clean printed image on a piece of paper, manufacturers formulate dyes and pigments in water to produce the image and add solvents to help the ink dry quickly. Resins, which are buffering agents to hold the ink at a constant pH, and surfactants (a type of detergent) enhance the ink mixture's flow properties. Ink formulas also contain a chemical preservative to keep the mixture stable over a long period of time and a range of temperatures. Many of these ingredients damage ecosystems by interfering with the reactions inside living cells.

Even without a hazardous material present, some computer components remain difficult to handle and recycle, a process called *e-cycling*. The plastic shells of many electronic devices present problems for recyclers because they contain an assortment of screws, plates, labels, paints, glass, and various coatings. E-waste in fact contains several hundred components. It is time-consuming to recover these components and it uses a large amount of energy in the process. As a result, they are usually thrown out. As the following table shows, electronics contain a very heterogeneous mixture of substances that waste managers confront every day.

The United States generates more than three million tons (2.7 million metric tons) of e-waste annually. The waste includes hundreds of thousands of miles of computer and telecommunications cables plus keyboards and computer mice. People rarely think about additional parts such

| MATERIALS OF CONCERN IN E-WASTE | | |
|---|---|---|
| **COMPONENT** | **HAZARD** | **EFFECT ON HUMAN OR ECOSYSTEM HEALTH** |
| nonhazardous metals, plastic shells, screens, cables | excess nondegradable bulk in landfills | loss of habitat for landfill |
| plastics | bromine-containing flame retardants | *bioaccumulation* in human and wildlife tissue |
| contacts, switches, and batteries | cadmium and nickel | cadmium *toxicity* in plants, wildlife, and humans; nickel allergies in humans |
| metal housings and joints | hexavalent chromium corrosion protector | toxicity in liver and kidneys; potentially carcinogenic |
| circuit boards and cathode ray tubes | lead | toxic to animal nervous systems; toxic to plants |
| flat screen displays | mercury | highly toxic compounds accumulate in food chains |
| wires and cables | polyvinyl chloride (PVC) | incineration creates toxic dioxins and furans |
| springs, relays, connectors, motherboards | beryllium | beryllium dusts are highly toxic to humans when inhaled |

as power plugs, chargers, routers, ports, and memory cards when they reflect on e-waste. Taken together, e-waste represents one of the most heterogeneous mixed waste categories in the world. Because waste from today's models of electronic devices creates an almost insurmountable problem, new approaches to e-waste management will be needed in the near future. Clean computers and new methods of computer distribution may help solve the problem; these topics are covered later in this chapter.

## HOUSEHOLD BATTERIES

Americans buy about 3 billion household batteries each year. The term *household battery* refers to any of the following battery styles: A, AA, AAA, C, and D series, 9 volt, and button style watch/calculator batteries. The most common of these are alkaline and zinc-carbon batteries, also referred to as heavy duty or general purpose batteries.

The purpose of all batteries is to convert chemical energy to electrical energy. To do this, today's batteries work on the same principles that Italian physicist Alessandro Volta developed in 1880. Within each battery, a chemical reaction forces electrons to move from one charged pole, the negative electrode (cathode), to an oppositely charged pole, the positive electrode (anode). The electron flow through a field of electrolytes (charged molecules) within the battery develops a current. The total voltage held by a single battery is determined by the type of metal inside it. For many years mercury served this purpose but this toxic metal has been replaced in newer batteries. For example, alkaline batteries rely on a zinc electrode at the anode and manganese oxide at the cathode. An alkaline, or basic (nonacidic), salt such as potassium hydroxide (KOH) serves as the electrolyte source. By contrast, general purpose batteries usually contain zinc and carbon electrodes within an acidic electrolyte. Alkaline or general purpose batteries work well in flashlights and radios, but more sophisticated electronics rely on stronger lithium-ion, nickel-metal hydride, nickel-cadmium, or other types of high voltage batteries. These batteries are also classified as household batteries.

Regardless of the type of household battery, when all the electrons have migrated from the negative to the positive electrode, no further chemical reaction can take place and the battery is said to have run dry or to be dead. Dead batteries can be thought of as small packages of metals that can contaminate the environment.

Each American household throws out about eight used batteries a year, totaling 2.5 billion nationally. Rechargeable batteries, which also contain heavy metals, delay this rate, but they too

## ELECTRONICS POLLUTION

Electronics pollute the environment in ways large and small and their varied array of hazardous components complicates the problem. In addition to the metals in circuitry and switches, harmless looking outer casings of phones, notebook computers, and video players contain an invisible hazard. The plastics and metal plates and hinges in electronics are treated with a group of chemicals called *protectants* that increases their durability.

eventually add to the waste. The battery industry has greatly reduced the hazards associated with household batteries so that they are safe to discard with nonhazardous household waste. Batteries in cell phones, computers, and most other advanced electronics, however, contain mercury, lithium, lead, or cadmium, and they should be brought to licensed recycling centers for safe disposal, even though battery recycling is not efficient, and the majority of household batteries sent to recyclers end up in municipal landfills. Recyclers handle mostly button batteries, like those in watches, and some nickel-cadmium varieties, and dump the rest in landfills. Millions of mercury-containing batteries remain in use, and these often end up in landfills where they represent a potential source of mercury pollution.

New battery technology focuses on two areas: alternatives to metals and biological batteries. Energy-conducting biomolecules—large compounds made by natural enzymes—may soon allow battery makers to eliminate the use of the toxic metals and corrosive acids and bases inside batteries. In the second focus area, biological batteries called *bio-cells* run on the energy-producing pathways of microbes. Bio-cell development will depend on the collaboration of scientists from diverse fields: biology, biochemistry, materials science, physics, and engineering. A Massachusetts Institute of Technology bioengineer Angela Belcher has been developing a new generation of batteries based on biological reactions. She explained to the *Forbes/Wolfe Nanotech Report* in 2003, "I learn about as much as I can about different fields. It is a way to ask questions and approach problems that are next generation problems." Perhaps batteries will soon be grown rather than built, and biological materials will serve a wider range of needs. In 2008 Belcher spoke with *Chemical and Engineering News* and added, "I think that multidisciplinary thinking and approaches are going to go a long way toward making major breakthroughs. I think that can be key to pushing science forward and solving the next generation of challenges, whether it's in energy, medicine, or the environment."

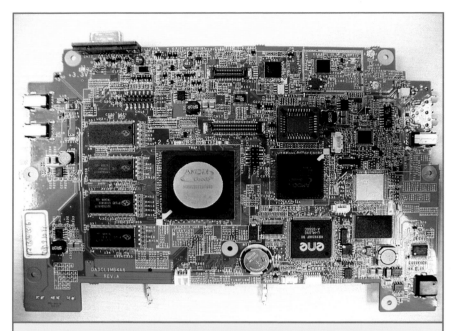

Computer circuit boards contain an array of nondegradable wastes that must be recovered, separated, and sent to an industry that can use them. Circuit board materials include, but are not limited to, toxic heavy metals, nonhazardous metals, alloys, plastics, composite materials, and coatings.

The protectant hexavalent chromium, for example, protects metal components from corrosion. Brominated flame retardants are another type of protectant applied to the inner and outer plastics of televisions, computers, and cables. These bromine-containing substances include mainly the following chemical groups: polybrominated diphenylethers, polybrominated biphenyls, and polychlorinated biphenyls (PCBs), all suspected of poisoning food chains by disrupting hormone function. There is an irony attached to today's protectants: They give new electronics a longer lifespan than previous models, but most devices are thrown away far before they reach the end of their usefulness.

## HEAVY METALS FROM E-WASTE

Mercury is but one of many heavy metals recovered from e-waste. Heavy metals are metals in Groups 3 through 12 in the periodic table of elements,

## White Goods

*W*hite goods is the term for a category of solid waste that is also a type of e-waste. The white goods category includes large household appliances—their standard color has traditionally been white—that are likely to contain hazardous chemicals or metals. Typical white goods are refrigerators, freezers, ranges, dishwashers, washing machines, clothes dryers, air conditioners, furnaces, and hot water heaters. Because of their bulk, white goods take up landfill space, and some landfill owners have therefore stopped accepting them. The components inside white goods have low resale value, which further decreases their worth as recyclable waste. These disadvantages make white goods an unattractive waste for general recyclers, so today a subset of recycling businesses specializes solely in these items.

The Steel Recycling Institute (URL: http://www.recycle-steel.org) states that a typical appliance is about 75 percent recyclable steel. White goods contain additional more hazardous materials that require special care during the breakdown of each discarded appliance. *Chlorofluorocarbon* (CFC) refrigerants in discarded appliances such as air conditioners and dehumidifiers require special handling according to the regulations set down by the Clean Air Act. Since the 1980s, scientists have known that CFCs destroy the atmosphere's ozone layer that protects Earth from ultraviolet radiation. For this reason, parts containing CFCs must be removed from appliances before the rest of the unit can be sold for reuse. To recycle CFC-containing refrigerants, white goods recyclers extract CFC-containing coils from each appliance without breaking them and spilling CFCs. Recyclers then transfer the liquid contents to safe containers. After sending the coils and the appliance's metal shell to a scrap metal facility, the recycler uses specialized equipment to remove impurities from the CFCs. The recycler incinerates the impurities and sends the CFC refrigerant to a chemical company that reconstitutes it to a reusable form.

CFCs are not the only hazard removed from white goods. Almost all white goods contain oils that can be recovered and reused and older appliances (models made before 1995) may contain mercury switches from which the mercury also can be recycled. Refrigerators and freezers made

*(continues)*

*(continued)*

before 2005 often contain insulating foams made of ozone-depleting substances. By managing these harmful substances, white goods recycling delivers the following benefits:

- controlled management of hazardous materials
- reduction of ozone-depleting emissions
- prevention in the release of chlorinated compounds, mercury, and oil
- reduction of materials entering landfills
- recovery of scrap metal and other recyclables

or put another way, heavy metals comprise the metals between copper and bismuth in the periodic table, and e-waste is a storehouse of these hazardous chemicals.

Numerous components within a single electronic device likely contain heavy metals. For instance, almost all circuit boards contain lead soldering that holds the conductors to the support board. Some boards also contain small amounts of gold as a binding material. Other heavy metals in e-waste include mercury in the circuitry of relays and switches and cadmium in resistors, semiconductors, and nickel-cadmium batteries. Cell phones and devices with liquid crystal displays contain the metals beryllium, lead, and cadmium in addition to arsenic. As mentioned previously, chromium shows up in discs and in flame retardants.

The sidebar on page 42 "Household Batteries" reviews the common elements in batteries. The major metals of concern in batteries are lead, cadmium, zinc, mercury, silver, and copper. Battery makers have reduced the amount of mercuric oxide by 86 percent in the past several years as well as the total number of mercury-containing batteries they sell, but batteries nonetheless account for almost one-third of mercury used worldwide.

Heavy metals are toxic to living cells at very low concentrations, and they accumulate in animal tissue, particularly the kidneys. Heavy metals do not break down in the environment so they accumulate in food webs.

Animals such as humans that reside at the top of food chains are at the greatest health risk from heavy metals due to bioaccumulation. In bioaccumulation, the concentration of a toxic substance increases at each step up the food chain. A typical example is provided by mercury that settles to the bottom of rivers. In the river's sediments the elemental mercury converts into an organic form called methylated mercury. This form of mercury enters billions of microbes in the water and the sediment, which are then ingested by thousands of microscopic invertebrates. As each food is eaten by an organism above it in the food chain, the mercury content disperses within smaller and smaller populations of animals, but at higher concentrations. The organisms at the very top of the food chain receive the highest amounts of mercury in their food and concentrate high amounts of the metal in their tissue.

Heavy metals cause harm to almost every metabolic system in the body. These elements and other metals such as lithium interfere with enzyme systems in plants and animals. In animals, they affect the kidneys and the nervous system the most, but also damage lungs, the heart, skeleton, and reproductive organs. Long-term exposure to heavy metal pollution is known to cause some cancers. In 2007, P. Parthasarathy, who is a recycling expert based in Bangalore, India, described to *Gulf News* the situation in his country, saying, "We have seen children waist-deep in cables, keyboards and circuit boards picking through a soup of hazardous chemicals, including lead and mercury, to find components or bits of metal they can sell." Serious health risks are occurring at this moment due to e-waste that has not been properly managed.

## SEPARATION AND REUSE

The first step in dealing with waste electronics involves dismantling them into their components in a process called *demanufacture*. Demanufacture and recycling of the recovered parts can be thought of, collectively, as the treatment method for e-waste. Five to 10 percent of discarded computers undergo this demanufacture and are then fitted with new drives, circuitry, and memory and graphics cards to turn them into reusable products. E-waste that is not rebuilt in this manner begins a long journey that may extend halfway around the globe.

Up to 80 percent of American e-waste heads to countries where low-wage workers, including children, demanufacture it more cheaply than can be done in the United States. These workers often lack protective clothing and are exposed all day to hazardous fumes and metals. Outside the plant, soils often contain dangerous levels of toxic chemicals. New Delhi, India, is one of many places serving the global need for disassembling computers, and it is beginning to receive worldwide attention due to the demanufacture industry's health risks. Tons of e-wastes from around the world arrive in India for burning, crushing, and even boiling. In his interview with the Middle East's *Gulf News,* P. Parthasarathy added, "We have seen (recyclers) breathing in dioxins as cables and casings burn around them."

Demanufacture wastes build up in ever-growing piles, many of which leak hazardous materials that had not been removed during the disassembly process. Since countries earn money by receiving as much e-waste as they can, they sometimes dump the wastes whole into open landfills without any dismantling. China, India, and Nigeria now struggle with massive stockpiles of whole e-waste. The Basel Action Network coordinator Jim Puckett warned in the *Washington Post* about the global toxic waste trade

E-waste recycling means manual dismantling of electronic devices into their components, many of which are hazardous. These workers in Africa take apart computers and printers without protective clothing, eye protection, or equipment to prevent inhaling hazardous vapors. Workers in the e-waste demanufacturing industry have increased health risks. *(Informationsdienst Wissenschaft)*

as far back as 2005, "The U.S. just looks the other way as we use these cheap and dirty dumping grounds."

Complete demanufacture done the proper way depends on a thorough separation of all the device's components. It is a labor-intensive and hazardous process. Workers wear protective coveralls and goggles to protect their eyes from broken chips of plastic and metal. They manually remove batteries, circuit boards, and switches and disassemble servers, keyboards, and other peripherals. Other workers sort large plastic pieces by polymer (the main material that makes up plastic) composition, bale them, and send the plastic to companies that melt and use them in new formulations. Smaller pieces go into grinders where they are broken into small shards and separated from metal fragments.

Scrap metal recyclers receive the metals from these operations and divide them into grades. For example, aluminum may be separated into light and heavy grades for different industrial uses. Recyclers also retrieve all the copper they can from wiring and use chemical reactions to recover it from alloy parts. All of the recovered metals go to smelting plants that turn the metals into industrial raw materials to be used for other purposes.

Glass from cathode ray model televisions, meanwhile, falls into categories based on barium, leaded, or mixed-grades. Each glass can then be used in making new cathode ray tubes. These specialized activities illustrate that as e-wastes continue to grow, recyclers must refine their expertise to make use of almost every material. Some e-waste recyclers now divide components and materials into as many as 50 categories of scrap.

Suppliers of industrial raw materials and the metals industry make up the biggest consumers of demanufactured e-waste. Computer makers have also started programs for reducing the postconsumer waste that comes from their products. Sometimes a state or a country's environmental laws spur manufacturers to be more efficient. Maine was the first state to hold computer manufacturers responsible for their obsolete and discarded products by requiring companies to start recycling programs. Additional governments now exert similar pressures on the electronics industry. Japan, South Korea, and many European countries have instituted so-called take-back laws to reverse the flow of e-waste back to its source. An entire new industry has begun to blossom based on this concept of reversing the flow of products at the ends of their lifespan. This field is called *reverse logistics*. The following sidebar "Case Study: Community Answers to Surplus Computers" describes a successful recycling computer program taking place on a local level.

## CASE STUDY: COMMUNITY ANSWERS TO SURPLUS COMPUTERS

nly 10 to 15 percent of electronics are recycled in the United States, so they may be considered one of the nation's recycling priorities. The U.S. Environmental Protection Agency (EPA) publishes a list of companies that accept e-waste for a fee and offers other resources for community e-cycling programs. People may not be inclined, however, to drive long distances to drop-off sites, and they may be even less motivated to haul heavy appliances. Many municipalities have addressed this obstacle by setting up periodic free e-waste collection days; some waste haulers now provide appliance pickup for a small fee.

Community-based organizations have added another option for e-wastes. Goodwill Industries in Austin, Texas, operates a state of the art demanufacturing plant that refurbishes more than 3,000 computers each year, then sells them in their store. At the same time, employees receive training in recycling and retail business. Pittsburgh, Pennsylvania's Goodwill offers in-home computer cleaning and upgrade service that keeps older models running longer while teaching information technology (IT) skills to its employees.

In California, Santa Barbara's nonprofit organization Computers for Families collects and rebuilds donated computers, keyboards, mice, and cables. Youths from county-managed programs refurbish about 50 donated computers each week. Their training includes how to upgrade, clean, test, and certify the rebuilt units, and the organization then provides the refurbished equipment to low-income households. As an added benefit, participants receive an opportunity to learn IT, computer installation and operation, and business skills. The organization's Matt Zuchowicz explained at its five-year milestone in 2002, "We believe that access to technology at home is critical to building the skills essential for success in the 21st century." As computer technology advances, students will also receive up-to-date training.

Programs like Computers for Families show the innovative ways people can delay putting e-waste into the environment. These programs serve families in need, and they teach meaningful skills to students from high-risk circumstances. Future goals may include expanding these services to handle other types of electronic items in addition to computers.

# NEW TECHNOLOGIES FOR
# REDUCING E-WASTE

The most effective way to reduce e-waste is to begin on the product's drawing board. Each new electronic product should consist of a plan for designing more efficient models and minimal packaging. Design engineers concentrate on the new model to be made while production engineers look for ways to achieve low-waste manufacturing. Project managers oversee the entire product life cycle from design to demanufacture, a path that can be quite complicated for today's electronics. Renee St. Denis led a team of such professionals in 2006 at Hewlett-Packard's Roseville, California, recycling plant; one of her group's objectives was to find ways to design lower waste-producing computers. Some managers call this start-to-finish process "Designing out the e-waste." Ms. St. Denis said in a 2006 *New York Times* article, "We want all these valuable resources put back into the economy in some way, shape or form." Some of the ideas already part of new designs include the following component substitutions:

- tin-silver-copper alloys for lead solder
- recycled polyethylene for polybrominated flame retardants
- recycled plastics for new plastics

On a regional level, the Northeast Recycling Council (NERC), representing 10 northeastern states from Delaware to Maine, oversees a program in which volunteer industries and the public make green choices in electronic products. NERC evaluates computer and monitor designs for their total *carbon footprint* (related to the amount of natural resources it consumes) and then offers tips on better designs, raw materials, energy efficiency, and product use. NERC's mission statement summarizes the organization's goals: "To advance an environmentally sustainable economy by promoting source and toxicity reduction, recycling, and the purchasing of environmentally preferable products and services."

On the West Coast, California has taken another approach by enacting e-waste laws that many people consider to be stricter than federal laws. In 2003, the state introduced the Electronic Waste Recycling Act, establishing funds for collecting and recycling certain e-wastes. This act covers, among other activities, the manner in which computer makers cooperate with stores to collect and recycle e-waste. The California Department of Toxic Substance Control's information officer Ron Baker explained in 2005, when the act was amended, "Someone starting from scratch can go to www.eRecycle.org to look at documents for ideas of how to do it, what we look for during inspections, what forms to fill out." California is now one of the most successful recycling states in the nation; on Earth Day 2007 Californians recycled 1 million pounds (454,000 kg) of electronic wastes—in a single day.

Some U.S. and European companies believe product life cycles are more easily controlled entirely by manufacturers and not by government, especially in areas related to costs. St. Denis remarked to *Recycling Today* in 2004 on the reservations the business community had at the time of the California bill's passage: "Because this is a state program, there is going to be state overhead, and there is overhead at several levels. There is going to be a lot of cost to this program that I think people didn't expect."

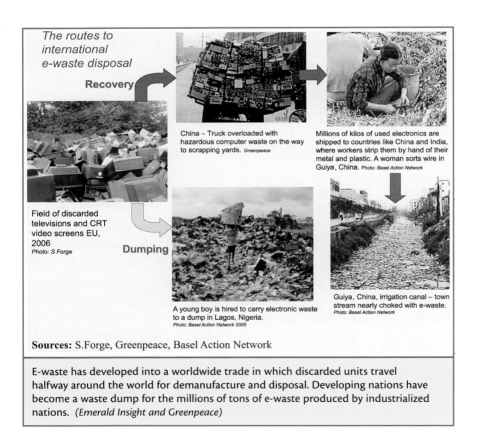

The routes to international e-waste disposal

Recovery

Field of discarded televisions and CRT video screens EU, 2006
Photo: S Forge

China – Truck overloaded with hazardous computer waste on the way to scrapping yards. *Greenpeace*

Millions of kilos of used electronics are shipped to countries like China and India, where workers strip them by hand of their metal and plastic. A woman sorts wire in Guiya, China. Photo: Basel Action Network

Dumping

A young boy is hired to carry electronic waste to a dump in Lagos, Nigeria.
Photo: Basel Action Network 2005

Guiya, China, irrigation canal – town stream nearly choked with e-waste.
Photo: Basel Action Network

**Sources:** S.Forge, Greenpeace, Basel Action Network

E-waste has developed into a worldwide trade in which discarded units travel halfway around the world for demanufacture and disposal. Developing nations have become a waste dump for the millions of tons of e-waste produced by industrialized nations. *(Emerald Insight and Greenpeace)*

Two other ideas have emerged recently as possible solutions to excess computers: (1) leasing and (2) new packaging. Leasing programs make computers available to consumers directly from the manufacturer, who earn money through leasing contracts rather than sales. This is not a new invention. For many years, telephones belonged to the local phone company and were not owned by customers as they are today. Efficient packaging also helps reduce computer-associated waste. Five options for achieving this goal show promise. First, companies are now moving toward instruction manuals accessed online rather than printed. Second, designing more durable electronics resistant to breaking extends products' life cycles and reduces the total amount of protective packaging. Third, designing smaller products reduces waste volume and packaging. Fourth, replacing polystyrene insulation (Styrofoam)—it is already banned in many U.S. cities and in Europe—with water-soluble or compostable fiber packaging reduces waste. Alternative packaging may be composed of starch-based plastics, shredded corrugated paper, or recycled Styrofoam. A newer packaging made of water-soluble sugar cane fiber reinforced with bamboo decomposes within 180 days. Last, manufacturers must control their urge to use extra packaging for advertising a product's features. This extra packaging often does little to protect the product itself.

## CLEAN COMPUTERS

Clean computers or green computers are models designed to contain energy sources, construction, and circuitry that reduce the entire computer's carbon footprint. Sustainability designers Dawn Danby and Jeremy Faludi noted, "Even with cheap energy, it's beginning to cost more to power a computer for four years than it costs to buy the computer." Batteries, liquid crystal display screens, lamps, and plastic parts will probably be the first components to be reengineered to make clean computers. Clean computers also require manufacturing processes that reduce the use of chlorinated organic solvents such as dichloromethane, methyltrichloride, trichloroethylene, tetrachloroethylene, and ozone-depleting chemicals. These solvents currently are used for plastics and polymer manufacture, engine cleaning, and general parts cleaning. Clean computers are indeed one of green technology's most promising areas for the future.

Two areas underway in clean computer development are new plastics and alternative energy sources. New plastics for computers consist of either biodegradable synthetic polymers or biomolecules, both of which must be strong enough to protect hard drives, circuitry, and other parts of electronics. Companies small and large have been trying to develop polymers that deliver all the characteristics needed by the electronics industry. One such firm, the DuPont Company, owns a technology called PDO, short for the compound 1,3-propanediol. PDO technology includes a corn sugar–based resin as an ingredient for a biological plastic called Bio-PDO, which provides two crucial characteristics: durability and strength. These plastics derived from a biological process are called bioplastics. The electronics and automotive industries have the greatest interest in bioplastics as a green raw material in their manufacturing processes. Dennis McGrew, the head of NatureWorks, explained to the *International Herald Tribune* in 2007, "As prices for fossil fuels soar and as the environment becomes an ever larger concern, ecofriendly plastics are becoming increasingly competitive, though they still remain a niche market." Green materials must overcome their own high costs before they become a viable choice as a raw material. The petroleum industry has argued that bioplastics may not be the answer because many will be difficult to degrade in nature. Judith Dunbar spoke for the American Chemistry Council in 2007 when she said, "It's not just bio-based versus petroleum-based. I believe conventional plastics would probably be better than renewables over a full life cycle." Bioplastics must overcome the stumbling blocks of cost as well as biodegradability.

Alternative energy sources have generated similar excitement in the business world. Universities and innovative companies have developed biological fuel cells, or bio-cells, as alternate energy sources for in-home and portable electronic devices. Algae and bacteria may soon be harnessed for generating a current that can run small electrical items. Oxford University chemistry professor Fraser Armstrong has been a pioneer in bio-cell development. He described the scientific and business potential of this new technology in a 2007 press release from the American Chemical Society: "The technology is immensely developable. We are at the tip of a large iceberg, with important consequences for the future, but there is still much to do before this generation of enzyme-based fuel cells becomes commercially viable." Bio-cells may one day supply the energy currently used by electronic devices in sleep mode or by backup power devices. Biological reactions also have promise as alternative semiconductors and for creating images on a screen to replace today's liquid-crystal screens. Some of the obstacles that must be hurdled include durability, enzyme activities, enzyme stability, and the potential expense of any catalyst needed to work with the bio-cell's enzyme system.

Convincing an entire industry to build clean computers may be no less difficult than forcing the automotive industry to produce cars that avoid fossil fuels. Businesses understandably resist any change that raises the cost of doing business. Therefore industry organizations and governments might offer incentives for electronics manufacturers to produce a more sustainable product. In order for clean computers to be as desirable for manufacturers as they are for consumers, they will need to attain some or all of the following attributes:

- user-friendly
- durable design for easy manufacture and demanufacture
- minimal packaging or packaging that serves as part of the device
- biological energy sources
- biodegradable plastic parts or bioplastics
- hardware-free joint connections
- halogen- and heavy metal–free components
- circuitry made of biochips (biological semiconductors)
- non–ozone-depleting protectants
- minimized or eliminated sleep modes

The EPA recently took on the task of converting its computer system to green models. EPA administrator Stephen L. Johnson stated in a 2008 press release, "Throughout the U.S., computer centers are becoming the fastest growing users of energy. By investing in energy efficiency in our own computer centers, we are proving that doing what's good for the environment is also good for the bottom line."

## CONCLUSION

E-waste is the fastest growing type of solid waste in the world. It includes computers and their supporting devices, telecommunications devices, electric appliances, and a wide range of other electronics. New models of electronics enter the market at a rapid rate, but only a small percentage gets recycled, leading to increasing amounts of e-waste worldwide.

E-waste treatment means the disassembling and recycling of products' components. The process of dismantling electronics and recovering reusable parts is called demanufacture. In demanufacture, workers remove toxic materials and send them to safe disposal, treatment, or reprocessing for reuse. Common hazardous materials that are part of e-waste are heavy metals, chlorofluorocarbons, oils, and toxic chemicals. Specific industries can reuse some of these hazardous substances but not all of them. For instance batteries containing toxic metals, acids, and bases make up a subset of e-waste that contributes to the worldwide tonnage of waste produced annually. Demanufacture also results in large amounts of nonhazardous components. These are separated into categories, and many are reprocessed, cleaned of impurities, and supplied to industries as raw materials.

The e-waste crisis demands that new computers be designed and manufactured to use less raw materials and produce less e-waste. Industry organizations and governments are working together to encourage manufacturers to take responsibility for reducing postconsumer waste. New manufacturing methods and new product-return programs will probably lead the way in tandem.

The next generation of computers will be clean computers. That is, they will be made largely through sustainable processes. The percentage of biodegradable ingredients in their structure and packaging will increase and their drain on nonrenewable resources will decrease. New materials and innovative designs are essential for these and other types of electronics. The electronics and appliance industries have formidable challenges ahead, but e-waste is an area in waste management with a great need for improvement.

# INCINERATION

Solving waste problems by burning may date back as far as the first human communities. Early civilizations could not ignore an inescapable fact of biology any more than people can today: If man finds a place to live on day one, then he must deal with his garbage on day two.

Open pits filled with refuse on the outskirts of early settlements were no doubt unpleasant, and they filled quickly. Burning probably took shape as the first innovation in the history of waste management, and for centuries burning remained the easiest way to get rid of a dwelling's waste. Waste simply disappeared in a puff of smoke.

In time, inventive individuals built incinerators to reduce several days' worth of waste to a tidy pile of ash. These metal cylinders reached temperatures of 1,600–2,550°F (870–1,400°C) and did the job faster than burning wastes in open pits. By the 1800s, entire cities relied on incinerators. In 1874, Nottingham, England, developed organized municipal trash pickup and routine incineration of each load. The United States built its first incinerator in 1885 on Governors Island in New York Harbor. Within the next 25 years, close to 200 more incinerators went into operation around the country, and the number reached 700 by the 1940s.

As piles burned down, heavy smoke and ash swirled upward. So too did unpleasant odors. Because waste disposal was largely a local concern, as it is today, national leaders paid little attention to towns choking on increasing amounts of hot airborne particles. At the same time that many towns installed new incinerators, others had had enough of the fouled air, and they abandoned or dismantled the contraptions. Many communities again turned toward the town dump for waste disposal.

Manufacturing expanded in America in the 1940s and '50s, and this growth would later have an impact on the business of incineration. Manufacturing plants began pouring millions of tons of pollutants into the air during this time, but the responsibility for keeping air clean fell on local authorities and not industry. The federal government had no involvement in air quality until the mid-1950s when the skies above Washington, D.C., began filling with pollutants blown in from industrial centers in the Midwest. A suddenly alarmed Congress responded by passing the Air Pollution Control Act of 1955, which called for local governments to set up pollution control for their communities. For a more thorough discussion, see "Case Study: The Development of the Clean Air Act on page 64. In the next decade and a half, however, few regions made significant headway toward complying with the air quality regulations. This notable failure plus the growing environmental awareness of the public prompted Congress to readdress the problem of air pollution. As often happens, industry's desire for growth conflicted with the public's desire for a clean environment. One mayor from a small industrial city put it this way during House debates on the amendments, "If you want this town to grow, it has got to stink." Manufacturing plants and their incinerators continued to belch heavy emissions. Industry would soon be pitted against environmentalists.

The young environmental movement, basking in good feelings generated by the first Earth Day on April 22, 1970, tackled air pollution in earnest. In 1970, Congress felt the public's frustration regarding air quality and enacted the Clean Air Act of 1970 to establish defined air quality *standards* for each state to follow and deadlines by which the states had to meet these standards. (Individual states found it difficult to meet the new standards, so Congress further amended the act in 1977 to extend the deadlines.) The amendments took into account the needs of some states that were home to large industrial complexes. Cleaning up the air would not be a simple task, particularly because in the 1970s analytical laboratories could not yet identify all of the various airborne pollutants that scientists are able to measure today. In President Richard Nixon's message to Congress in 1971, he said, "While we still have a long way to go before we meet our ultimate objectives, it is important to emphasize that we are making substantial progress. For example, there is evidence that the air in many of our cities is becoming less polluted, although the data [are] still incomplete. Total emissions from automobiles and the use of persistent pesticides are going down. On the other hand, there is no basis for complacency, as the level of total pollutants in our environment is still rising." With that message in mind, many people surely turned their attention to one troublesome source of pollutants: incinerators. The new air quality law led to the shutdown of hundreds of incinerators that had been spewing pollutants for decades, and by the end of the Clean Air Act's first year of enactment less than 70 incinerators remained in operation.

Though fewer incinerators operated at hundreds of factories, industrial waste kept mounting. Some factories tried to make old incinerators comply with the air pollution law by installing *scrubbers*—devices to trap pollutants flowing up smokestacks—but many other manufacturing plants simply stopped incinerating their waste and looked for other disposal methods to avoid the costs of complying with the new law. Landfills seemed to be the easiest solution, but the country's remaining landfills were nearing capacity. The time had come to fix incineration's flaws.

On May 3, 1994, the U.S. Supreme Court ruled that incinerator emissions and ash were to be treated as hazardous waste and landfills could no longer serve as an inexpensive alternative to installing a clean-burning incinerator. *New York Times* writer Linda Greenhouse reported, "In a decision that could substantially increase the cost of waste, the Supreme Court ruled today that any toxic residue created by burning household and industrial waste in municipal incinerators must be treated as hazardous waste and not dumped in ordinary landfills." The court also ruled that certain wastes must be treated only by incineration while meeting the new air quality standard. Some towns abandoned plans for building new incinerators because of the costs of retrofitting old equipment to meet new requirements or the cost of handling waste in order to stay within the dictates of the ruling. Lawyer Karen Florini of the Environmental Defense Fund explained the new requirements: "The main source of dangerous chemicals in municipal trash is discarded batteries and electronic equipment, which can be separated from the trash before incineration."

As the costs of complying with stricter environmental laws increased, U.S. industries followed a path that countries in Europe had already explored: *waste-to-energy* (WTE) incinerators. WTE incinerators played a role in sustainable communities like that in Kalundborg, Denmark. In the 1990s, the United States had a number of new WTE plans on the drawing board, but it lagged behind other countries in building them. Today, more than 1,000 waste-processing plants worldwide generate steam for producing heat or electricity. Japan and Switzerland treat 70 percent of their municipal solid waste (MSW) in incinerators connected to energy-capturing systems. France has shunned incineration since 1998 when dioxins from three large incinerators showed up in food webs, but it now draws energy from 36 percent of the MSW it treats. The United States burns only 15 percent of MSW in modern incinerators, Canada merely 8 percent. These low percentages reflect the fact that in North America, people do not welcome any type of incinerator. The environmentalist Ellen Connett spoke about her concerns over the role incinerators play in waste management in a 2007 radio interview: "We are asking our engineers the wrong questions: how do we get rid of waste. What we should be asking our engineers is, how can we stop making waste. With all the packaging we have, if we can't reuse it, recycle or compost it, we shouldn't be making it." In other words, Connett feels incinerators have no role in today's waste management. Hundreds of new incineration plants have been delayed or cancelled because of the NIMBY phenomenon—Not In My Back Yard.

Incineration remains an important method of waste treatment in the United States even with challenges to overcome. Incineration's future success depends on clean technologies so a skeptical public will accept it. This chapter examines new technologies that make incineration a cleaner way to tackle the global problem of too much waste.

## METHODS IN MUNICIPAL WASTE INCINERATION

Incinerators destroy hazardous and nonhazardous wastes by heating them to 1,600–2,200°F (870–1,200°C); some models go as high as 3,000°F (1,600°C). An oxidizing agent added to each load helps all the organic matter combust completely to carbon monoxide, carbon dioxide, and nitrogen gases, plus water and a small amount of hydrochloric acid. Incinerators work well for destroying hazardous organic compounds such as cyanides and sulfides as well as heavy metal–containing materials. Incinerators do not completely destroy the metals, but they reduce the combustible wastes to ash. This incinerator ash is now classified as a hazardous waste and only landfills made for hazardous wastes can accept ash. New thermal (high heat) methods have also helped reduce the amount of ash in landfills. One such method is vitrification, discussed in the next chapter.

Waste experts are beginning to find new uses for the ash left over from nonhazardous waste incineration. In Europe, companies make solid blocks from the ash for use in construction or road building. Other entrepreneurs mix ash with soil and sell the mixture as landfill cover or landscaping material. Incinerator ash mixed with cement or other hardening materials may soon be tried for building artificial reefs in places where natural marine habitats have been damaged.

It can be difficult to dismiss the image of incinerator smokestacks billowing dangerous emissions. But ash emissions today create less pollution than in years past because new cleaner incinerators release far lower amounts of toxic gases than old incinerators. Still, some gases escape even from new models. Incinerator-produced sulfur dioxide and nitrogen oxide combine with moisture in the atmosphere to make sulfuric acid and nitric acid, respectively, and the mixture creates acid rain. Another

Incineration has progressed from a simple waste-burning method that yields ash to innovative treatments that destroy all waste in high-temperature processes. Advanced thermal destruction methods may soon reduce organic waste and medical waste to their basic elements and simple gases. *(iStockPhoto)*

potential disadvantage of incineration relates to the efficiency of combustion. Complete combustion depends on a good supply of air. Without ample amounts of oxygen, heating becomes inefficient and organic wastes cannot break down completely to gas and water. The partial reactions inside the incinerator then release toxic pollutants such as dioxins. Lastly, incineration plants are saddled with the disadvantage of busy truck traffic carrying waste loads in and treated ash out.

Despite the drawbacks, incineration reduces the amount of waste that must go to a landfill by 70 to 90 percent. Incinerators treat large waste loads so quickly that wastes hardly have a chance to build up. Incinerators also take up much less space than landfills and they are easy to operate compared with more advanced technologies in waste treatment.

Improved incineration technology includes equipment designed to control air turbulence inside the combustion chamber. This helps give the thermal reactions excess air, which leads to high-efficiency combustion. Operators also carefully control the high temperatures in new-model incinerators. At an appropriate temperature for each load, wastes decompose in a matter of seconds. The time period in which waste loads undergo treatment is called residence time, and incineration offers one of the fastest residence times of all waste treatment technologies.

Chapter 1 showed the ways in which different types of wastes have very different compositions. For that reason new models of incinerators have been designed for treating different types of waste loads. Hospitals use incinerators of a size and capacity to treat their medical wastes, which can be largely solid materials. These materials are described in more detail in the sidebar "Hospital Waste" on page 72. By contrast, wastewater treatment plants need units that dispose of tons of semisolid sludge. The common incinerators used today are multiple hearth, cement kiln, rotary kiln, and fluidized bed, but there are many more, listed in the following table. Fluidized bed incinerators are small units (9–25 feet in diameter [2.7–7.6 m]) that run at lower temperatures than standard incinerators and work best on moist sludge from wastewater treatment. By contrast, multiple hearth incinerators consist of a more complex design and run at high temperatures to destroy dried sludge from wastewater and chemical treatment plants. Incinerators running about 1,800°F (1,000°C) are best for wood wastes.

| COMMON INCINERATION PROCESSES | |
| --- | --- |
| TRADITIONAL INCINERATION PROCESSES | NEW INCINERATION PROCESSES OR MIXED PROCESSES |
| cement kiln | recirculating fluidized bed |
| rotary kiln | auger combustor (starved air combustion) |
| multiple hearth | two stage (starved air combustion) |
| fume | liquid injection |
| fluidized bed | molten salt |
| industrial boiler | oxygen enriched |
| multiple chamber | catalytic combustion |
| cyclonic (low air combustion) | infrared |

Incinerators large or small and at every temperature need a good air supply for combustion to progress. Some incinerators have air inflows designed to push air over the heating unit. This serves two purposes: (1) it helps combustion, and (2) it helps decrease particle emissions.

Japan, Norway, and Sweden prefer fluidized bed incineration because it is highly efficient (combustion is complete) and it emits lower levels of nitrogen oxides, sulfur dioxides, and dioxin. In fluidized bed incineration, equipment disperses each waste load onto a moving bed of limestone and sand, which is heated by an upflow of air from a furnace below the bed. The limestone neutralizes acids and minimizes acid gas release. The one minor disadvantage of fluidized bed incineration is a larger amount of ash produced at the end of combustion.

## INCINERATION AND ENERGY PRODUCTION

Energy sustainability occurs when energy released by one process is captured for use in another process. These WTE operations are also called

(continues on page 66)

## CASE STUDY: THE DEVELOPMENT OF THE CLEAN AIR ACT

Between 1920 and the 1950s, industrial emissions in and around London had blocked the city's sunshine by 20 percent. Londoners coined a new term to describe the noxious mix of smoke and fog, *smog*. A retrospective by BBC News on the "Great Smog of London" stated, "At Sadler's Wells, the opera *La Traviata* had to be abandoned after the first act because the theatre was so full of smog." In the United States, air quality had deteriorated almost as badly and the hazard continued to grow for the next four decades. In 1999, Clean Air Task Force Technical Coordinator Conrad Schneider summed up the nation's smog problem: "Despite popular impressions, this is not just a Northeast problem. From Texas to Illinois from Georgia to Maine, and everywhere in between, people are admitted to the hospital for serious, prolonged respiratory distress due to ozone smog." The federal government's passage of the Air Pollution Control Act of 1955 and the Clean Air Act and its amendments in the 1970s attacked air pollution from two directions: by fixing the already dirty air and by preventing new sources of air pollution. Then as now, air quality solutions focus on three problem areas: (1) urbanization, (2) industrialization, and (3) large numbers of vehicles on the roads.

The Air Pollution Control Act of 1955 provided money for research into air pollution. However, the Clean Air Act of 1963 was the first federal law mandating levels of allowable emissions in the air, and the levels would be based on scientific findings. Since individual states had the job of figuring out how to abide by the law, each state developed a plan based on its air quality. These situations, of course, varied from state to state. The clean air regulations in New Jersey's manufacturing centers were quite different from those in Montana. State-by-state regulations pointed up another simple fact: Air moves. As an example, rural Pennsylvania would likely have to contend with pollutants from Cleveland, Chicago, or Detroit or even farther away.

In 1970, Congress amended the act to allow the federal government to set limits on air pollutants for the entire country, and the U.S. Environmental Protection Agency (EPA) accepted responsibility for enforcing the revised law and for getting all states to comply with the new standards by 1975. To this day, several texts refer to 1970's Clean Air Act as the first law passed in this country for preventing air pollution. Amendments to the law later authorized the EPA to improve the standards for air quality, meaning the agency could set limits on the amounts of certain substances in the air. From the start, states with large industrial centers struggled with the law.

Most states failed to meet the 1975 deadline, which became extended to 1977, although over the next decade more and more regions began reducing dangerous emissions. A formidable barrier to success remained, however, because of scant information on the types and amounts of emissions that posed the greatest health threats. It would be hard to set limits on a compound if the compound was unknown. Analytical equipment improved quickly during this period and

Cities such as London, England, have suffered smog events that have permanently affected residents' health and the behavior of native fauna. London's most famous smog problems occurred in the 1950s, but this photo taken in 2005 shows that some air pollution problems persist despite new antipollution laws and technologies. *(iStockPhoto)*

became capable of finding things other than particles and incinerator gases. New equipment detected acid rain, smog, and toxic volatile chemicals, even ozone gas, and in 1990 Congress amended the act to include these additional pollutants. More important perhaps, the act put tighter controls on the sources of air pollution and also required that polluters use the best available technology to clean up their emissions.

The 1990 amendment expanded the list of hazardous air pollutants to almost 200 and allowed for the addition of new pollutants. It also took into account state-by-state differences in geography, population, and industry. Today, states may adopt regulations that fit their circumstances, but they may not weaken the federal laws. California has led the way in enacting state air pollution laws that are significantly stricter than the regulations of the federal Clean

*(continues)*

*(continued)*

Air Act. Much of the opposition to air quality standards has come from industries, particularly the automotive industry. Says energy expert and economist Nancy Ryan, "Because California is the only state that is permitted to implement stricter pollution controls under the Clean Air Act, it is in a unique position to influence automakers and pave the way for other states to follow." California had been granted a waiver from the Clean Air Act's requirements because the state already had a stricter air pollution program before the federal act was passed.

The EPA provides the public with updated information on the nation's biggest polluters, and it also provides companies with information on specific chemicals released into the air and guidelines for reducing their release. Rather than produce a complicated plan covering hundreds of pollutants, the EPA has selected six different pollutants as benchmarks for ambient air quality. Ambient air is the air in a person's near surroundings, meaning the air people breathe. States monitor the levels of the following six pollutants for an overall assessment of air quality: carbon monoxide (CO), nitrogen dioxide ($NO_2$), sulfur dioxide ($SO_2$), ozone ($O_3$), lead, and particles that contain heavy metals.

Since the Clean Air Act of 1970, ambient air levels of the six priority pollutants have decreased almost 50 percent. Smog levels have not dropped, however, due to continually high emissions from industry and vehicles. The act's complete success may occur only when these emissions have been drastically cut, but this has proved to be a difficult challenge.

*(continued from page 63)*

EFW, for energy-from-waste. Despite incineration's bad reputation as a polluter, it possesses characteristics that make it suitable for WTE.

Incineration has high energy costs, meaning it consumes large amounts of energy to run the incinerator. There is, after all, a price to pay for destroying thousands of tons of matter in a few seconds. The energy escapes as heat through the incinerator's walls, into hot ash, or up the smokestack. In order to make incineration an energy-sustainable process, those energy losses must be captured for reuse.

From 1980 to 1990, WTE plants treating waste in the United States doubled to almost 100 and by 2002 the number had reached several hundred. WTE growth has since slowed. In 2005, about 33.4 million tons (30.3 million metric tons) of MSW were combusted with some energy recovery, but this amounted to only 14 percent of the country's total MSW load. Perhaps some communities oppose any type of incineration based on a history of incinerator emissions causing severe health hazards and even deaths before the Clean Air Act took effect. The environmental *think tank* Environmental Literacy Council summarized on its Web site (URL: http://www.enviroliteracy.org) the current state of incineration in this country: "Initially, incinerators were used simply to reduce the volume of waste. Now, most are WTE facilities which use the combustion process to also generate useful by-products, including heat, steam and electricity." The organization also points out the obstacles: "Despite their long history, the use of incinerators continues to be controversial due to issues, such as the emission of gaseous pollutants. Despite the use of pollution control devices, there is also concern over escaping ash particles that may contain trace quantities of heavy metals, dioxins, and other substances." WTE proponents must convince the public that WTE is an important way to build sustainability as Kalundborg, Denmark, has successfully done.

Three different types of WTE incinerators capture thermal energy: (1) mass burn, (2) modular, and (3) refuse-derived. Each can treat 100–3,000 tons (91–2,722 metric tons) of MSW a day, which produces the equivalent heat energy from burning 25–750 pounds (11–340 kg) of coal. A mass burn incinerator is a large unit fed by overhead cranes. Mass burn incinerators handle bulky wastes that resist combustion by other heat treatments. Fans attached to mass burn units blow in air to help combustion, then a water jet cools the hot ash as it exits the combustion chamber. Each ton of waste treated by mass burn can produce up to 500 pounds (227 kg) of ash ready to be delivered directly to a landfill. Some waste-treatment sites run a conveyor belt directly from the incinerator to a landfill for easy disposal of the ash. Modular incinerators are small units that serve well for on-site waste treatment at places like hospitals (see the sidebar "Hospital Waste" on page 72). The heat given off by modular and mass burn WTE incinerators boils water into steam during the combustion of each waste load. A pump pushes the steam into distribution lines that supply heat for buildings or power turbines to generate electricity. The third type, refuse-derived fuel incinerators, reduces waste into pellets. Each pellet contains an energy value similar to that of coal, and, in fact, a small amount of coal is sometimes added to the pellets to boost their energy value.

What if *all* the waste entering an incineration plant was destroyed and nothing but energy resulted? A new type of incineration called *plasma arc* technology may achieve this 100 percent conversion of waste to energy. The National Aeronautics and Space Administration (NASA) developed plasma arc technology in the 1960s, but it has been applied to MSW treatment only within the past decade. Countries in Europe, principally Germany, the United Kingdom, Ireland, Canada, India, and Japan, currently operate plasma arc facilities.

Plasma arc technology vaporizes waste, explaining why it is sometimes called plasma arc vaporization or *gasification*. Plasma arc behaves as a closed-loop system; gas emissions do not escape but rather go to a gas-powered turbine. The turbine's energy output is then redirected to one of two forms of power: (1) steam to run the waste treatment facility, or (2) electricity for neighboring homes and businesses. By using combustible gas emissions that come from the incineration, the final emissions contain almost no hazards. Japan's plasma arc incinerators, for instance, produce emissions far below the country's allowable emission levels, standards much stricter than those in the United States. St. Lucie County in Florida is in the planning stage for a plasma arc facility for treating the county's waste and generating power. Power utility officials hope to vaporize 1,000 tons (907 metric tons) of household garbage a day and increase the amount within five years to 3,000 tons (2,422 metric tons) a day. In the process, they will generate electricity for about 40,000 homes. Mark McCain, a manager at the Florida Municipal Power Agency, said in 2008, "If this project works as we hope it will, this will be a renewable source of energy which is something that many of our customers are encouraging utilities to do today."

Plasma is a form of matter created when solids are heated to temperatures like that found on the Sun's surface. Plasma arc technology combines electricity and high pressure—similar to conditions that produce lightning—to create plasma from solid waste. At 10,000°F (5,540°C),

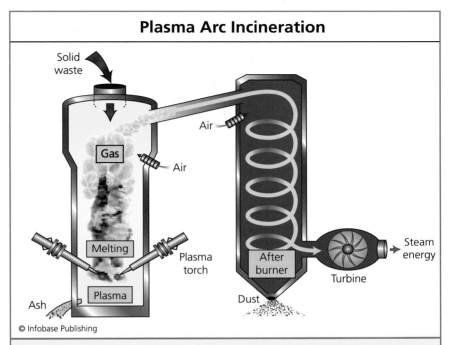

## Plasma Arc Incineration

Solid waste

Gas

Air

Air

Melting

Plasma torch

Plasma

Ash

After burner

Dust

Turbine

Steam energy

© Infobase Publishing

Plasma arc technology is an extreme high-temperature treatment that destroys all waste and turns it into a gas within seconds.

almost every type of waste turns to gas in seconds. Florida officials expect the site to produce 120 megawatts of electricity daily from an output of 80,000 pounds (36,287 kg) of steam when the plant reaches full capacity. The steam in turn will be sent to energy-producing turbines at a nearby fruit juice producer and a small amount of metal-containing ash will be sold to construction firms as a raw material. County officials also hope to treat wastes now sitting in the area's landfills, and the county estimates that the new technology may in 20 years remove the landfill wastes that have been piling up since the 1980s. County Commissioner Chris Craft explained in 2006 to the Associated Press, "It addresses two of the world's largest problems—to deal with solid waste and the energy needs of our communities. This is the end of the rainbow. It will change the world." Even if plasma arc technology's expectations fall short, this method should contribute in a sustainable manner to solving the growing waste problem.

# INCINERATED MATERIALS AND AIR QUALITY

Incineration plants in the United States have since the Clean Air Act's passage made adaptations to remove substances from their emissions and to stay within legal limits for air pollutants. All incineration plants conduct trial runs—monitored by the EPA—to assure their pollution controls work properly before they begin operations. During this trial, incinerator operators collect data on each pollutant and compare the emission levels to legal limits. Scrubbers do most of the work in removing pollutants from the emissions, but new air pollution devices such as advanced filters, electrostatic precipitators, and afterburners will soon help.

Incineration plants, factories, and hospitals install scrubbers to remove particles called *fly ash* and toxic compounds from flue gases. As dirty air flows through a scrubber, the air passes into a cylinder that swirls it through a substance meant to collect the pollutants before the air exits the smokestack. Two types of scrubbers do this: wet scrubbers and dry scrubbers. Wet scrubbers use water or chemical spray to collect hazardous materials. Water offers low cost and it circulates well through the system. Dry scrubbers use alkaline powders to absorb acidic gases and filters to

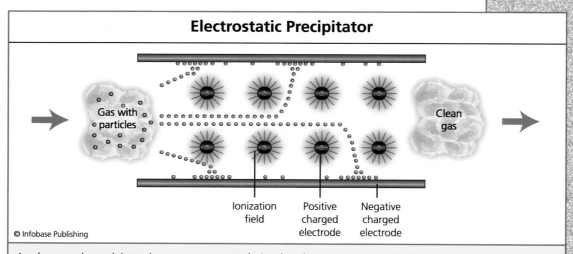

**Electrostatic Precipitator**

Gas with particles

Clean gas

Ionization field

Positive charged electrode

Negative charged electrode

© Infobase Publishing

An electrostatic precipitator is a waste treatment device that cleans contaminated gas by putting an electrical charge on hazardous particles and then removes those charged particles by applying an electric field of opposite charge.

capture the particles. Scrubbers now remove particles of less than 0.4 μm in diameter.

Cloud chambers follow the principles used by scrubbers, but with an innovative approach. A cloud chamber contains positively charged water droplets that attract tiny negatively charged and some neutral particles. After a period of time to allow this collection to take place, the electrical charge inside the chamber reverses. The cloud chamber, already filled with negatively charged droplets, captures positive particles and any remaining neutral ones. All the particles then condense in the chamber's water, safe for removal.

Filters provide a simple solution for cleaning pollution out of the exhaust flowing from incinerators. Pollution control filters consist of durable fabrics, fiberglass, or carbon-based materials. All types of filters pass tests for strength and resistance to high temperatures, and they usually contain other benefits, such as flame retardation. Modern filter materials also withstand strong acids and bases and other caustic chemicals, are unaffected by reactive chemicals, and do not readily age or disintegrate.

Innovations in filter fibers allow this technology to become ever more efficient in pollution control. New filters contain fiberglass, polypropylene, polyester, acrylic, polyphenylene sulfide, or polytetrafluoroethylene fibers. Carbon fibers make up another filter category that acts by trapping particles and adsorbing molecules, meaning they cause pollutants to adhere to carbon fiber's outer surface. (Adsorption is any adherence of molecules to the outer surface of a material; absorption is the uptake of molecules into a material.) Each carbon fiber contains thousands of filaments wrapped into a tight bundle, which collects organic compounds, inert substances, and small particles. Carbon beds, by contrast, provide a layer of activated carbon rather than bundles. Activated carbon is any natural source of carbon (i.e., coconut shells and wood chips) that has been pulverized to increase surface area and so increase adsorptive capacity. When incinerator exhaust passes through a bed of activated carbon, the carbon removes pollutants from the exhaust similar to the way a paper filter removes particles from dirty water. The carbon's tiny pores admit pollutants and hold onto them by either chemical bonding or less specific electrostatic attractions, called van der Waals forces. The pulverized material can be formed into pellets, granules, or powder. After a period of time, activated carbon fills with pollutants and must be replaced.

Filter technology had once been limited to removing only fine, solid particles from liquids. Newer technology targets volatile compounds in addition to solid particles. For instance, some fabrics and fibers trap aero-

## HOSPITAL WASTE

ospital waste is also known as medical waste or infectious waste and is sometimes nick-named *red bag waste* because technicians store and decontaminate these wastes within red or orange biohazard bags. Almost 80 percent of hospital waste resembles household trash, but the remaining 20 percent contains hazardous substances: infectious microbes, toxic sub-stances, and radioactive materials. In addition to *pathogens*, hospital waste contains noninfec-tious matter such as radioisotopes, chemicals used in diagnosis, and therapy drugs. The *World Health Organization* (WHO) groups medical wastes as follows:

- infectious wastes—microbial cultures, wastes from infected patients, blood, discarded diagnosis samples, infected laboratory animals, bedding, contaminated swabs and bandages, and contaminated devices such as disposable probes and dental instruments
- anatomic wastes—body organs and parts, and animal carcasses
- sharps—syringe needles, disposable scalpels, and blades
- chemicals—solvents and disinfectants
- pharmaceuticals—expired, unused, or contaminated drugs, vaccines, and sera
- genotoxic wastes—drugs, usually for cancer treatment, that are mutagenic, tera-togenic, carcinogenic, or otherwise hazardous
- radioactive wastes—containers contaminated with radioactive remnants and radiochemicals used in therapies
- heavy metal wastes—broken thermometers and discarded equipment that may contain mercury or other heavy metals

Infectious and anatomic wastes make up 15 percent of hospital waste, and the sharps cate-gory is growing. Each of these wastes comes not only from hospitals but also from outpatient clin-ics, laboratories, blood banks, nursing homes, mortuaries, dental offices, and veterinary clinics.

Vaccination programs carried out on a global scale contribute to the fast growth of the sharps category within hospital waste. (The medical community gives about 12 million injections annu-

solized calcium-, chlorine-, or sulfur-containing organic compounds and particles containing metals. The specific chemical's properties determine how it will be captured. Cadmium, lead, and arsenic, for example, disperse

ally and not all syringes and needles are disposed of properly.) To understand the magnitude of this problem, the Coalition for Safe Community Needle Disposal (URL: http://safeneedledisposal. org) reports that 3 billion injections take place each year in the United States by in-home syringe users. This excludes hospital and doctor's office injections. Rather than solve a health problem, improperly or illegally handled vaccination wastes spread infection when people and animals scavenge on the disposal sites. Each disease outbreak has the potential to add enormous amounts of injection wastes into the environment. A 2001 measles outbreak in western Africa illustrates why sharps have become a concern: That single outbreak produced 660 tons (599 metric tons) of injection waste.

Most U.S. hospitals operate their own incinerators to destroy infectious microbes in waste loads. A portion of hospital waste contains plastic fluid bags and blood bags, which have for a long time been made of polyvinyl chloride (PVC). An inefficient hospital incinerator—one working at lower than normal temperatures—creates intermediate compounds such as dioxins, furans, and polychlorinated biphenyls (PCB) from incomplete combustion of PVCs. Dioxins, furans, and PCBs cause potentially serious health problems when inhaled by people, and presumably by wildlife as well. The WHO Web site (URL: http://www.who.int/) provides guidelines on incinerator design, operation, maintenance, repair, and operator training, and it promotes new cleaner technologies for destroying hospital wastes.

Veterinary medical waste carts hold infectious materials needing decontamination, usually by incineration. Most medical waste is nonhazardous, but hazardous components hold pathogens, chemicals, and radioactive materials. The World Health Organization cites medical waste as one of the most poorly managed categories of hazardous waste due to poor regulations and enforcement. *(Cornell University College of Veterinary Sciences)*

into small water droplets within incinerator exhaust, and the filter then catches them. Other metals such as aluminum, copper, and iron do not vaporize, and they remain in the incinerator's ash.

Fabric filters designed for metal pollutants work for 18 to 36 months before they must be cleaned. Workers may simply shake the filter bag to do the job or in other cases backwash the filter, which means reversing the airflow through the filter to dislodge ash. Both fabric filters and carbon filters, therefore, require maintenance to work their best at cleaning emissions.

Electrostatic precipitation offers a more advanced method of cleaning emissions. Similar to the cloud chamber, a precipitator attached to an incinerator's outflow pipe uses chemical forces to remove pollutants. Charged plates inside the precipitator give a negative charge to particles as they flow through the first section of the chamber. The particles then pass through positively charged plates that pull up to 99 percent of toxic emissions from the exhaust flow. Precipitators are less efficient in treating emissions that do not have an electrical charge, such as the emissions from burning low-sulfur coal. Another disadvantage of precipitators is that they can create a fire hazard if unburned gases should enter the device and ignite.

An afterburner is yet another technology for controlling pollution from incinerators. Modeled after devices used for boosting thrust in military jets, afterburners combust the air that escapes an incinerator's burning process. It recycles emissions through a second combustion, which further heats and expands the gases. Afterburner technology may provide another advantage in the future: the capture of re-burned gas conversion to electricity.

## CLEAN INCINERATION

Clean incineration destroys waste while producing no harmful emissions. Scrubbers, filters, and afterburners each can help make a traditional incinerator work as a clean incinerator. Advanced clean technologies now strive to attain sustainability by capturing the energy given off from clean incinerators during the waste treatment process and reusing it to treat more loads. Plasma arc technology delivers this type of potential benefit because it eliminates an entire waste load and converts it into a form of energy that can be reused by a community. Clean and sustainable incineration may recover energy in the form of heat, light, gas, or steam. For example, steam heat generated in combustion powers a turbine, which converts it to electricity. The electricity then powers the incineration plant and any excess electricity enters the community's power grid for use by local households and businesses.

Scrubbers and related devices have helped in the development of clean incinerators, but this technology is still in its early stages. Plasma arc incinerators may be the first to attain clean treatment, although they currently have high installation and operation costs. A method called *refuse-derived fuel incineration* involves steps before the actual treatment process, mainly presorting each waste load to remove glass, metals, and other noncombustibles. This adds cost, but presorting creates a more homogeneous waste load than nonsorted waste and thus the WTE process becomes more efficient. Costs have become such an overriding concern in the waste industry—cost is a major concern in almost every other industry as well—that a disturbing trend is beginning to take shape. Some communities have redesigned their WTE plants back to the old style of incineration because they believe they will save money. The excess heat energy generated during the waste treatment dissipates and is lost forever.

## RISK ASSESSMENT AND GLOBAL NEEDS

The United States must depend on more than one method to treat the 250 million tons (227 metric tons) of MSW it generates annually. Countries such as Japan and Switzerland have learned to generate energy from at least 50 percent of their wastes, but other countries struggle with huge amounts of waste and little money for new technologies. (The United States falls somewhere in the middle of these two extremes.) Open-air burning and old-style incineration still predominate in much of the world for treatment of wastes.

Since the early 2000s, Germany has developed a unique approach in waste management. German waste incineration firms now import raw waste from other countries that do not have the money or the inclination to treat their wastes in a clean manner. In fact, Germany has become a major importer, perhaps the biggest importer, of hazardous wastes. Its neighbor, the Netherlands, has closed down two of its own incinerators in Rotterdam and now sends the loads to Germany. In 2007, the German magazine *Der Spiegel* quoted Paul Braams of Rotterdam's waste combustion service, "[Germany's] got such good facilities, why should we spend good money to bring our own incinerators up to date?"

Is incineration at a crossroads? It seems to be an all-purpose treatment method as shown in the following table. While certain parts of Europe and Japan have embraced WTE technology, the United States has not committed to it mainly because communities dislike incinerators in general. German leaders and waste industry representatives used to find incineration an objectionable way to treat wastes, but the profits they have reaped from their new tactic have reversed their thinking. In the 2007 *Der Spiegel* article, the country's environmental minister Sigmar Gabriel reasoned, "With its very good facilities for incinerating hazardous waste, Germany is assuming a part of the general environmental responsibility." Other German citizens have predictably questioned the safety and merits of importing waste to make money. Any incomplete combustion processes will only pour dangerous chemicals into German skies, while it may or may not solve a waste problem in another part of the world. Johannes Remmel, the secretary of the Green Party in Düsseldorf's parliament, told *Der Spiegel* he was concerned as to "whether it's the job of local waste disposers to acquire hazardous waste from all over the world."

The United States and other countries continue to question whether to put more resources into incineration technology or to abandon it alto-

| TYPES OF WASTE TREATED BY INCINERATION | |
| --- | --- |
| TREATED ALMOST EXCLUSIVELY BY INCINERATION | INCINERATION PLAYS A SUPPORT ROLE IN TOTAL WASTE TREATMENT |
| MSW | excess pulp and paper |
| wood scraps, furniture | yard trimmings |
| wastewater sludge | plastics |
| hazardous chemicals and solvents | agricultural wastes (crop cuttings) |
| hazardous by-products of other cleanup/treatment methods | |
| plants used in phytoextraction | |

gether and turn to different treatment methods. Areas outside North America certainly have serious air pollution that may be further harmed unless they convert to clean incineration. Parts of Taiwan, China, and India, for example, often exceed 500 μg of particles per 35 cubic feet (1 m³) of air; levels of more than 100 μg threaten human health. If incinerators continue to run, new clean technologies and WTE are the best hopes for incineration's future.

# CONCLUSION

Burning trash is one of the oldest waste disposal methods, and incinerators play the major role in the thermal, or heat, treatment of hazardous and nonhazardous wastes. Incineration has fallen into and out of favor as new technologies develop and are perceived to be less damaging to the environment. But the sheer volume of the world's waste demands a variety of treatment methods and incineration remains one reasonable approach. Incineration's main benefit lies in its capacity to reduce large volumes of waste in seconds to a small amount of ash. But incineration also presents the disadvantage of hazardous emissions that incinerators release during combustion. Simple devices such as scrubbers and filters have reduced dangerous emissions from today's incinerators, but room for improvement persists.

The Clean Air Act of 1970 set standards for the amount and type of emissions that come from industries, vehicles, and incinerators. This law contributed to advances in building more efficient and clean-burning incinerators that convert waste to energy. A variety of models now serve specific waste producers, and advances will continue in the areas of cleaner incineration to destroy entire waste loads.

Clean incineration and WTE may become the next generation in incineration. Clean incineration remains in an experimental stage with plasma arc technology being the most promising innovation. Plasma arc incinerators operate in several countries outside the United States and may grow in number if their cost decreases. WTE has come closer than clean incineration to reaching its goals. Several countries now derive energy from more than 50 percent of their hazardous and nonhazardous wastes; the United States converts only 14 percent of its wastes to energy. Incinerators continue to suffer from a bad perception in the United States, particularly because of older models that produced dangerous emissions prior to the Clean Air Act. Incineration is nevertheless useful for treating large waste loads quickly and can play a major role in handling the large and growing volumes of municipal waste. Because of this potential, the technology of combustion must achieve truly clean thermal methods. In addition, governments and the public must learn to accept WTE plants if they desire sustainable communities in their lifetime.

# VITRIFICATION

Vitrification is a waste treatment that uses high temperatures to imbed hazardous waste in glass. The waste industry often refers to vitrification as glassification. Vitrification combines one of the oldest waste treatments, heating, with modern materials science. Materials science is the study of matter's chemistry and physical properties for use in new technologies. In vitrification, the chemical and physical properties of glass make it an excellent permanent receptacle for hazardous waste. This is possible because glass when heated becomes molten and in this form waste matter can be mixed into it. Cooling glass returns to a solid, which does not react with compounds in the environment. The solid glass blocks or logs produced by the vitrification process also safeguard against leaching of hazardous wastes into soil and groundwater. The attractiveness of vitrification rests in its ability to put waste into a permanently immobilized and nonreactive form.

Vitrification is one of four thermal technologies in waste treatment. Like incineration, plasma arc technology, and pyrolysis, it relies on an intense and controlled heating step. (Pyrolysis is a process that decomposes organic wastes by heating in the absence of oxygen.) Unlike the other three thermal methods, however, vitrification does not reduce the total waste load's volume, which is an obvious disadvantage of this technology. Vitrification equipment may be either stationary buildings or mobile units, so hazardous wastes may either be transported to a treatment facility or treated at the pollution site. On-site treatment offers an advantage when dealing with difficult-to-handle wastes or highly radioactive wastes.

This chapter examines the status of vitrification technology in today's waste management. It discusses the attributes of vitrification and its challenges. The chapter also reviews the history of vitrification technology,

Vitrification, or glassification, immobilizes radioactive wastes in a solid glass block. The wastes and glass must be mixed together when molten, as demonstrated in this picture. Vitrification may solve the legacy waste problem, meaning it will treat the stockpile of radioactive wastes produced many years ago in addition to the wastes being produced today. *(Knowledge 2008)*

its current uses and future advances, and the unique properties that make glass an attractive choice for storing hazardous wastes.

## HISTORY OF VITRIFICATION

In 1995, the U.S. National Research Council (NRC) called together international scientists to discuss the latest available technologies in radioactive waste management. Vitrification had been developed in 1978 for a waste treatment plant in Marcoule, France, but the unique process had at the time stirred little interest. Nuclear wastes in the United States continued accumulating, and the government could not postpone searching for a safe and secure permanent storage for its radioactive wastes. In 1989, *U.S. News and World Report* writer Michael Satchell noted, "Cleaning up radioactive and chemical waste at the nation's nuclear weapons plants and military installations looms as the biggest, toughest and most expensive task of ecological restoration in American history. It presents technical challenges equal to the Apollo moon landing and space shuttle programs, and it will cost roughly as much . . ." Beginning in the 1990s, two experimental sites, South Carolina's Savannah River facility and the West Valley Project in New York, invested in the innovative method of using glass as an inert storage form for large amounts of radioactive waste.

Early trials in vitrification at Savannah River and in New York went well, and other sites in the country considered the groundbreaking technology. In 1996, a committee organized by the NRC to study the role of vitrification as the next generation of waste disposal proclaimed, "Vitrification of high-level radioactive waste has received greater attention, worldwide, than any other high-level waste solidification process." Plans for new vitrification plants accelerated, especially in places with very large hazardous waste stockpiles. The undertaking failed to match the enthusiasm, however, and planners learned that vitrification is a highly technical process requiring well-trained operators and, therefore, very expensive. Communities hoping for a new vitrification plant learned that the science was costly and the facilities needed massive amounts of energy to melt the glass and mix it with wastes. Many lofty plans for new plants around the United States were probably scuttled just a few months after being drawn up.

The former Soviet Union and France meanwhile continued exploring vitrification technology for their nuclear wastes. The Soviets had built two plants in the 1990s using phosphate glass as a waste storage form for *high-level radioactive waste* (HLRW). During the same period, France expanded its vitrification capacity to handle industrial levels of waste, using *borosilicate glass* as the storage matrix. In addition to HLRW, French facilities planned to treat *low-level radioactive waste* (LLRW), such as medical and research materials, as well as the heavy metal–laden ash from incinerators. Alain Damien was a research scientist on one of France's vitrification projects at the time. He explained to *WasteAge* magazine, "We are looking to vitrify fly ash to make it inert and inoffensive." So inert, in fact, that a French company now turns vitrified fly ash and asbestos waste into a nonhazardous material for use in construction.

The U.S. waste industry remained slow in accepting this type of waste solidification. One question, yet to be answered, centered on the number of years radioactive wastes could safely remain in glass. Highly radioactive materials require safe storage for at least 10,000 years, and a few isotopes need a much longer period to decay to a less dangerous element. Plutonium-239—the fuel used in making nuclear weapons—has a *half-life* of 24,000 years, and the United States needs to find a secure, long-term storage for its large P-239 stockpile. The P-239 problem has been further complicated by stalled plans to use a storage site in the Nevada desert, Yucca Mountain. This site has been planned since 1985 by the U.S. Department of Energy (DOE) as a permanent underground storage for 34 tons (31 metric tons) of P-239. The Yucca Mountain repository has, however, been delayed numerous times due to the technical challenges of assuring safe underground storage for HLRW, opposition from local communities, and high costs.

*Deep burial* of radioactive wastes may be the safest approach for long-term disposal, but the Yucca Mountain project has shown that this method has many obstacles. Deep-sea burial and burial within polar ice sheets have been proposed as alternative solutions to aboveground storage. Deep-sea burial may be vulnerable, however, to volcanic activity under the ocean, which might crack open storage containers and release massive contamination. Also, in the past decade, polar ice caps have become vulnerable to melting due to global warming. With all these problems, the thought of launching the world's nuclear waste into space has been pondered, but huge expenses and accidents like that of the space shuttle *Challenger* in 1986 highlighted the enormous challenges of that option. Vitrification therefore remains today's truly innovative means of hazardous waste disposal accompanied by a long-term storage solution.

The United States opened its first vitrification plant in Savannah River in 1996. The large facility situated on a 310-square-mile (803 km²) site began working to reduce the nation's nuclear waste backlog. By 2005, the plant had made 7 million pounds (3.2 million kg) of radioactive glass. Across the continent in Washington, the Pacific Northwest National Laboratory (PNNL) in Hanford began gearing up its own larger vitrification operation. The site known simply as Hanford already processes more than 65 million gallons (246 million l) of nuclear weapons waste from military stockpiles. It also deals with a mixture of radioactive waste that had been indiscriminately dumped on its land for decades. These materials are described in more detail in the "High-Level Radioactive Waste" sidebar that follows. Later in this chapter, the sidebar "Case Study: The Hanford Nuclear Waste Site" examines Hanford in more detail. S. K. Sundaram is a materials scientist at PNNL who has summarized the technology's potential: "As scientists, we have a responsibility to find solutions to the 'legacy' waste problem confronting present and future generations. Vitrification is an ideal technology for immobilizing the wastes at Hanford and elsewhere. It is adaptable to changes in the composition of the waste while still offering the desired waste form properties."

The U.S. nuclear industry takes about 7,800 spent nuclear fuel assemblies out of use each year; about 2,200 tons (1,996 metric tons) of these

During the 1950s, the Savannah River site began to produce materials used in nuclear weapons, primarily tritium ($^3$H) and plutonium-239. Savannah River eventually built five reactors to produce nuclear materials. *(Department of Energy)*

assemblies hold very high levels of radioactivity. (The Nuclear Regulatory Commission [NRC] provides a list of the *spent fuel* locations in the United States on its Web page at URL: http://www.nrc.gov/waste/spent-fuel-storage/locations.html.) The waste needs a storage location unthreatening to nearby communities and the environment, and it must remain safely in place for thousands of years. The NRC has estimated that the nation's capacity to store spent fuel is 80 percent filled and could reach 100 percent capacity in less than 10 years. Vitrification is a 30-year-old technology, but because of the looming nuclear waste crisis it has become new again in a search for treatment solutions. In the near future, the public as well as the

## HIGH-LEVEL RADIOACTIVE WASTE

High-level radioactive waste consists of spent fuels from nuclear reactors and materials from military nuclear weapon production. HLRW is the most dangerous of all radioactive waste categories because of its high radioactivity and because it remains dangerous for thousands to millions of years. The DOE classifies radioactive wastes into four groups: high-level wastes, transuranic wastes (TRU), uranium mill tailings, and low-level wastes. The federal government is responsible for disposal of HLRW, which comes from nuclear power plants, naval vessels, obsolete nuclear weapons, and spent nuclear fuel reprocessing.

Spent fuel from power reactors consists of solid irradiated uranium oxide pellets enclosed within stainless steel tubes called *fuel rods*. Once a nuclear fuel has been used up, it is called *spent fuel* and is highly radioactive. Used fuel rods are extremely hot at the time of their removal from a reactor, and so they must immediately be cooled in either one of two kinds of storage casks: dry or wet. Aboveground dry casks made of concrete measure about the size of a school bus and provide an effective barrier against radioactive emissions. Wet storage casks, by contrast, contain fuel rods submerged in at least 20 feet (6 m) of cooling water. Despite their hazardous contents, spent fuel rods stored in casks do not explode and they do not begin to burn spontaneously as many people fear. After fuel rods have cooled, they must be disposed of in a safe manner, so they do not contaminate the environment. Today the country holds more than 100 stockpiles of spent fuel rods that wait for a secure storage site to hold them for thousands of years. For this reason vitrification disposal of HLRW may soon take on greater importance.

A small percentage of TRU has high radioactivity and is handled in the same manner as HLRW. Transuranic elements are those with atomic numbers greater than 92, that is, their atoms have more protons than a uranium atom. TRU also contains items of lower radioactivity—the wastes

U.S. nuclear industry may count on vitrification as the answer to disposing of radioactive wastes.

## THE PROPERTIES OF GLASS

What makes glass and glasslike compounds desirable for immobilizing massive amounts of society's most deadly wastes? Phoenician civilizations living in what is now Lebanon and Syria may have first developed glass as early as 3000 B.C.E. By heating and mixing the silicon dioxide ($SiO_2$) found in quartz with sand, soda ash (sodium carbonate), or limestone (calcium

discarded from nuclear weapons handling. Protective clothing, goggles, and contaminated debris and equipment provide examples of TRU wastes.

Before 1970, nuclear facilities put most of their HLRW into barrels and buried them beneath the ocean floor. TRU disposal had an equally casual ending, usually in shallow landfills. Now radioactive waste stockpiles sit at various nuclear facilities around the country in addition to a large and undefined amount of radioactive wastes in old dump sites. Vitrification serves two objectives in today's radioactive waste management. The first objective is to reduce the stockpiles of HLRW stored at the nation's nuclear facilities. The second objective aims to eliminate the unknown and often hidden amounts of radioactive wastes contaminating the environment, once they have been found.

Uranium mill and mine tailings represent other radioactive wastes that need permanent disposal and may eventually be included with HLRW. Uranium mill tailings consist of the sandy residues of uranium extraction from rock and ores. Tailings have traditionally been discharged as a semisolid slurry into holding ponds where they may languish for decades. Uranium mine tailings are slightly less hazardous than mill tailings. These consist of excess contaminated rock and ores that pile up outside uranium mines. For many years uranium tailings were poorly monitored for leaching into the environment, and today mill and mine tailings have fouled local soils and groundwaters. These by-products of uranium manufacture need safe underground disposal. Deep (2,150 feet [655 m]) underground salt formations have been considered for tailings and TRU, although this storage method has never been used on a large scale. Most likely, tailings and TRU will join the queue of materials waiting for treatment by vitrification.

carbonate), a hard glazelike material formed as it cooled. Glass merchants and artisans shaped the molten, pliable material into an infinite variety of forms. Early societies often used glass as a protective coating on the outside of water or wine containers. Successive generations made glass fit their practical and artistic needs. Native American tribes fashioned it into beads and ornaments as a type of currency, and, by mixing in metals, glassmakers gave glass a variety of colors. At the start of the 20th century, chemical companies began experimenting with new materials and new uses for familiar materials. Glass offered the following unique features: a hard yet transparent material; inexact molecular structure; pliability when heated; and ability to hold its shape indefinitely. Small companies that had perfected their glassmaking specialties found an opportunity to apply their expertise to new sciences. Glassworks eventually evolved from a small and oftentimes artistic specialty into a process for making a major industrial raw material.

Glass does not react biologically, so it is stable in the environment and helps keep wastes immobilized. Once glass has solidified, it withstands a high degree of compression and is not damaged by high temperatures. In addition the structure of glass is not exact; its crystal conformation varies depending on composition, and this too makes it useful for vitrification because it can accommodate a variety of chemicals. Most new vitrification methods have adapted France's choice of borosilicate glass, because it withstands extreme temperature ranges better than standard glass. (Other compounds dissolve or vaporize in extreme heat, but glass merely melts.) Borosilicate glass contains about 70 percent silicon dioxide, 5 percent boron oxide, plus lesser amounts of sodium and aluminum. Since silicon dioxide, or silica, and aluminum are the two most abundant elements on Earth, glass is perhaps the least expensive component of the entire vitrification procedure. Borosilicate glass is also an impermeable material; it does not allow water or other liquids to flow through it. Finally, like standard glass, borosilicate glass presents neither immediate health hazards nor any known long-term health problems such as cancer.

Despite glass's advantages, scientists cannot be sure of how vitrified glass will react in nature over thousands of years. Studies must be conducted on this question in addition to the long-term stress patterns and corrosion characteristics of glass containing a high concentration of radioactive materials. Some areas of study in glass chemistry for the purpose of improving vitrification are the following: corrosion reactions at

Glass has properties suitable for the safe storage of hazardous materials: inert, nonconducting, and stable. These vitrified glass bowls show the ease with which various materials mix with molten glass and then become immobile and stable within the glass matrix. *(Hectarus)*

the glass-air surface; silica loss from the glass crystals over time; and the solubility of specific compounds in glass.

Both the Hanford and the Savannah River sites focus on developing more efficient glasses that melt 20 percent faster than standard vitrifying glass. They call this new form fast glass, and it is capable of incorporating more total waste and in a shorter time than borosilicate glass. Speed in waste management helps reduce costs and may also help communities tolerate a new treatment technology in their midst.

## HANDLING IMMOBILIZED WASTE

Marie Curie's laboratory studies on radioactive substances laid the foundation for today's nuclear science. She died of cancer at the young age of 47 after 20 years of exposure to radioactivity from radium and other

*(continues on page 90)*

## CASE STUDY: THE HANFORD NUCLEAR WASTE SITE

The production of military nuclear weapons in the United States began in the 1940s during World War II with the Manhattan Project. This program recruited physicists, nuclear scientists, and engineers for the purpose of harnessing nuclear reactions for the world's most powerful bomb. One of the primary sites where the scientific studies took place is in a 560-square-mile (1,450 km$^2$) area in sparsely populated eastern Washington. By 1944, this site near the town of Hanford had become the world's largest nuclear production operation and held more plutonium than any other single country. By the end of World War II, Hanford had produced more than half of the country's weapons-grade plutonium. The properties of plutonium are described in the sidebar "Enriched Uranium and Plutonium" on page 95. Unfortunately, much had yet to be learned on the dangers of nuclear wastes in the environment and in the 1940s ecosystem was not yet part of everyday conversation.

From 1944 to the 1980s, Hanford produced more than 50 tons (45 metric tons) of plutonium. Over this period, indiscriminately dumped wastes filled the land, poured into lagoons, or sat in corroded storage tanks and drums. This poor waste management continued into the 1970s by which time the site had undergone significant environmental damage and posed a serious risk. A 1976 *New York Times* article reported, "A decade-old mixture of radioactive wastes blew up for unexplained chemical reasons today in a small building on the Hanford Nuclear Reservation, slightly injuring one workman and contaminating him and nine other workmen with radioactivity."

Radioactive iodine, I-131, used for reprocessing spent nuclear fuel makes up one of hundreds of different hazards in Hanford's waste mixture. From World War II through 1957, this material had been leaching into the ground while I-131–contaminated gases and particles drifted into the air and fell onto nearby homes, open waters, crops, and animals. Meanwhile, a heterogeneous radioactive mixture trickled from corroded storage containers toward the Columbia River and eventually made their way to the Pacific Ocean 200 miles away.

An even more disturbing occurrence took place from 1944 through 1957 as Hanford technicians measured the spread of radioactivity far outside the facility's property. The historian Michele Gerber searched the nuclear facility's records 40 years later and made a shocking discovery. "They just enlarged their sampling circles," she said, "to 25, 50, 100, 150 miles, all the way to Spokane and Walla Walla. Why didn't you [the Hanford facility] stop? Why didn't you change the production process to reduce the emissions?" By altering their sampling area, the Hanford technicians showed the levels of I-131 to be at safe levels in the environment. Gerber published *On the Home Front* in 1992 to detail the findings she had made on Hanford's activities. Hanford geohydrologist Roy Gephart responded to Gerber's information, saying, "The public felt betrayed. I talked to workers who said they were taken by surprise when they read the book. They felt deceived. That's why many Americans distrust us today."

In 1988, Congress mandated the Hanford Thyroid Disease Study (HTDS), an epidemiology study to determine who had been harmed by the emissions. Epidemiology is the science of determining the origins of disease outbreaks. Researchers first located people living downwind of the site and most likely to have been exposed to the hazard, then built detailed health histories for each person. Study personnel examined 3,500 people for evidence of thyroid disease and, after reviewing their data, concluded that I-131 pollution had not increased the incidence of illness in Hanford. The public immediately questioned the results, and many mistrusted any type of government-sponsored study. A professor of environmental studies at Portland State University, Rudi Nussbaum, declared, "The HTDS is a worthless study." The Centers for Disease Control and Prevention (CDC) has tried to assure the public of the meaning of the study results. In a 2002 report entitled *Summary of the Hanford Thyroid Disease Study,* the CDC stated, "The findings do not prove that Hanford radiation had no effect on the health of the area population. However, they show that if there is an increased risk of thyroid disease from exposure to Hanford's iodine-131, it is probably too small to observe using the best epidemiological methods available." In 2006, the CDC conducted a second study that showed people who had been exposed to Hanford's emissions long ago do have a slightly higher risk of developing thyroid disease. Trisha Pritikin lived in the Hanford area during the contamination and expressed her bitterness to Portland's independent newspaper, the *NewStandard,* in 2005 when she said, "Right now people like me are very disheartened and disillusioned by a government that told us everything was safe at Hanford . . . We sacrificed our health for the cold war." A Portland association called Hanford Watch now monitors Hanford's cleanup and keeps the public informed of all human health and environmental issues.

The Hanford site still contains a large volume of compounds of unknown identity, acids and other corrosive compounds, cyanide, nitrate, and chromium, in addition to radioactive chemicals that have had years to seep into soils and groundwaters. John Vienna, a chemist from the Pacific Northwest National Laboratory, has stated that "the Hanford wastes contain every element in the periodic table."

Hanford represents probably one of the world's most complicated waste cleanup projects, and its vitrification operations will become the world's largest, if those plans ever develop. Before treatment begins, acidic wastes must first be neutralized with millions of gallons of sodium hydroxide. Hazmat technicians then determine the different materials that have leached to

*(continues)*

(continued)

different depths in the land under the site. Layers, called gradients, containing different compositions must be removed and handled separately. Workers divide these gradients as best they can to separate nonradioactive from radioactive wastes and high-level from low-level radioactive materials. These gradient separations reduce the total volume of material to be vitrified and also remove any chemicals that interfere with the blending of waste into molten glass.

The federal government has set a deadline of 2028 for Hanford's cleanup. Hanford's wastes have been slated to go to the Yucca Mountain repository in Nevada, but construction delays at that facility have given Hanford, like other HLRW sites, an uncertain future. The vitrified material may well stay at Hanford for a long time. In 2004, the *Seattle Times* quoted Gerald Pollet, who was the head of a Hanford cleanup watchdog group, as reasoning, "I believe if Yucca Mountain is not safe, it shouldn't be open. Glassified, high-level waste should stay at Hanford. That's the safest thing."

(continued from page 87)

radioactive elements. Her daughter, Irène Joliot-Curie, followed in Marie's research footsteps and fell victim to cancer at age 59. The scientific community made little connection between radiation exposure and health during the time the Curies conducted their studies. But in the years following the Curies' work, hundreds of men who had spent careers in uranium mines began dying of lung cancer. Workers in industries that used radium and other radioactive materials also developed cancers. Not until the NRC formed in 1975 did a government agency set limits on the levels of radioactivity exposure for those working in nuclear science. Before then, the nuclear power industry policed itself on safety and the limits of radioactive exposure its employees received. More than one critic of the young nuclear industry complained that this scenario was akin to "the fox guarding the henhouse."

The NRC today balances a diverse mission of ensuring sound construction of nuclear facilities, controlling the movement of radioactive materials across the country and across its borders, and ensuring public safety from radioactive materials. Nuclear industry workers must have the proper safeguards against contamination. The NRC's Web site (URL: http://www.nrc.gov/) provides radiation workers with information in the following four areas:

1. dose limits for radiation exposure
2. the dose history database for each individual
3. retrieval of individual dose history reports
4. training requirements

Anyone handling radioactive wastes first receives training in the basics of nuclear science, meaning the study of the atom. This includes managers of radioactive waste streams and workers who handle equipment that touch nuclear materials or radioactive wastes. Technicians in vitrification plants and hazmat workers at waste sites contaminated with radioactive materials also receive this type of training. Training usually covers two subject areas: radiation and radioactive materials handling. Radiation is electromagnetic energy released from some forms of matter when the matter emits a decay particle. Each type of decay particle comes from an atom called a radioisotope. Four major types of decay particles escape from radioisotopes: protons (alpha particles), electrons (beta particles), gamma rays, and neutrons. Because a radioisotope is an unstable form of an element, it naturally seeks

## Types of Radiation

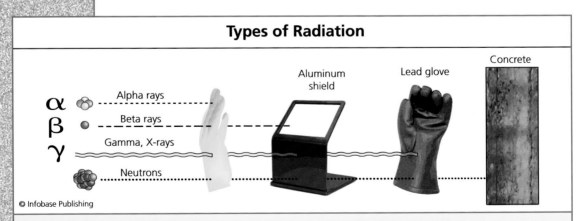

© Infobase Publishing

Before people understood the health hazards of radioactive elements, they handled materials such as uranium with their hands and without body or eye protection. Scientists today greatly reduce exposure to radioactivity by using simple materials that block radioactive emissions.

The level of protection hazmat workers require depends on the type of hazardous waste exposure. This flash suit is used by U.S. Air Force firefighters. The suit has an outer aluminized shell and can be decontaminated of hazardous chemicals, as shown in the background.

to regain its more stable form. When a radioisotope goes from unstable to stable, it releases energy in the form of radiation. Exposure to high doses of decay particle emissions, or radioactivity, is a health hazard. For that reason the second training topic covers the safe and proper handling of radioactive materials. Proper handling involves the use of protective clothing and physical barriers to keep radioactive particles from hitting a person's skin. Workers in nuclear facilities wear protective coveralls made of materials that block the type of radiation in their facility. Protection includes hoods with faceplates, gloves, and boots; shields also cover all snaps and zippers to prevent skin exposure.

Waste-handling facilities contain construction materials that provide a second line of defense against radiation exposure, called *secondary barriers*. Secondary barriers for HLRW consist of concrete with specific metals mixed in, and, in fact, facilities usually provide multiple secondary barriers for HLRW storage. These multiple barriers—at least one contains concrete—block neutron emissions, which are able to penetrate almost all types of matter. A dense layer of concrete serves as one of the few materials that stops neutron emissions. Most TRU contain the alpha particle emitters uranium, plutonium, neptunium, americium, curium, or californium. Alpha particles do not penetrate most materials, and barriers as simple as a sheet of paper block them. For this reason one barrier, usually metal, protects workers from most TRU. About 3 percent of TRU emit gamma radiation that penetrates more barriers than alpha particles. TRU that emits gamma rays behaves like HLRW because the rays penetrate biological and nonbiological matter. This explains why the DOE has classified these forms of TRU along with HLRW. For high-level wastes, lead, concrete, or steel provide the best protective barriers.

Safe handling also includes detectors measuring the amount of radiation exposure a person receives each month. In nuclear facilities, detectors throughout the building and on each person's protective clothing collect data on the dose of exposure to radioactivity. In the United States, the NRC works with the DOE to determine the limits of radiation exposure a person can receive. In 1928, the International Commission on Radiological Protection and the U.S. National Council on Radiation Protection and Measurement set a scale of radiation doses that a body can receive without harming the person's health. The scale indicates that different parts of the body may receive different dose levels and remain safe. People who work with materials must not receive more than five rems in a single year; this level is called the *total effective dose equivalent* (TEDE). (Rem stands for Roentgen Equivalent Man and is a unit of radiation dose that takes into account its effects on living tissue. Another unit called the sievert is becoming a more common measure of TEDE than the rem. One sievert equals 100 rem.) Another way of stating the U.S. regulations is as follows: The regulations allow workers to be exposed to radiation that can cause one additional cancer fatality per 400 workers per year. Workers who receive more than the TEDE must avoid radiation exposure for a period of time, depending on the amount of exposure, the type of exposure (inhalation, ingestion, etc.), and the organs exposed. Vitrification plants treat the three most dangerous levels of radioactive waste, described in the following table, so workers in these facilities require the highest levels of protection.

| TYPES OF RADIOACTIVE WASTE | |
|---|---|
| **TYPE** | **DESCRIPTION** |
| high-level (HLRW) | spent nuclear reactor fuel or wastes from reprocessing spent fuel |
| spent nuclear fuel | reactor fuel that can no longer sustain a chain reaction |
| transuranic (TRU) | by-products of spent fuel reprocessing, uranium mill tailings, contaminated industrial or research waste |
| low-level (LLRW) | any waste that is not the three above or naturally occurring radioactivity |
| mixed low-level | radioactive waste mixed with toxic chemicals |
| contaminated media | soil, sediment, water, or sludge with enough radioactivity to require special cleanup methods |

## INNOVATIONS FOR VITRIFYING RADIOACTIVE WASTE

Vitrification presents challenges for the waste industry compared with incineration, which is still the more common thermal waste treatment method. Vitrification operators manage large amounts of highly dangerous chemicals in each waste load. This treatment method also requires an enormous amount of heat to melt the glass and incorporate waste matter into it. Vitrification technology today involves the technical aspects of heating large amounts of glass and dangerous wastes to very high temperatures. By comparison, incineration seems simple: Load a furnace and close its door.

Despite vitrification's promising future, it has drawbacks that must be resolved. First, glass cooled too quickly solidifies before it should inside the heating chamber and stops the flow of materials inside the chamber. Such a backup increases the cost of the vitrification process and requires

## ENRICHED URANIUM AND PLUTONIUM

Two elements that make radioactive waste dangerous are uranium and plutonium. Uranium, named for the planet Uranus, is the heaviest natural element (atomic number 92) and the first element shown to be radioactive. French chemist Eugène-Melchior Péligot isolated and identified uranium in 1841, and 55 years later French physicist Antoine-Henri Becquerel demonstrated its radioactivity. For their hallmark studies on radioactivity, Becquerel and Pierre and Marie Curie were awarded the 1903 Nobel Prize in physics.

Uranium occurs naturally in minerals and is usually present as one of three isotopes: U-238 is predominant, and U-235 and U-234 occur in much smaller amounts. U-238 is the form used today in energy-producing nuclear reactors. Its long half-life (4.6 billion years) corresponds to low radioactivity because few atoms disintegrate at any given time. In the 1930s, German physicists discovered that bombarding U-238 with neutrons caused the release of additional neutrons and energy, a process called fission. They next learned that by controlling the released neutrons and re-bombarding the remaining nuclei, they could create a self-sustaining reaction. In 1942, the Italian scientist Enrico Fermi at the University of Chicago carried out the first such fission reaction, and by doing so he laid the groundwork for the development of modern nuclear reactors. The United States now operates more than 100 nuclear reactors for energy production, as do countries in Europe, Japan, and India.

Enriched uranium contains a blend of U-238 and U-235 and works in nuclear reactors that produce either energy or fuel for military weapons. Different grades of enriched uranium pertain to the amount of U-235 present in the mixture, from less than 1 percent to 90 percent (high grade enriched uranium). During a nuclear reaction, enriched uranium converts back to 100 percent of the U-238 form, and it is this form that makes up radioactive wastes from nuclear reactors.

Plutonium is not found in nature. Enriched uranium bombarded with deuterons—elements composed of one proton and one neutron—serves as the source of plutonium in the nuclear industry. Discovered in 1941, plutonium is named after Pluto, the dwarf planet. Since the discovery

*(continues)*

*(continued)*

of plutonium, 15 additional radioisotopes from this element have been found. One isotope, Pu-239, is much more active than U-238—it has a half-life of 24,400 years—and it fissions when it is bombarded with neutrons. Nuclear reactors called breeder reactors produce several thousand pounds of Pu-239 per year, which go into the making of nuclear weapons rather than energy production.

A second plutonium isotope, Pu-238, has a half-life of 87 years and emits only alpha particles. Energy from alpha particle emissions can be converted into heat, and then thermoelectric devices convert this heat into electricity; each gram of Pu-238 produces one half-watt of power. Today Pu-238 powers the generators aboard unmanned spacecraft and interplanetary probes.

Uranium and plutonium each play important roles in society, but society also must manage these dangerous elements in a responsible manner.

time-consuming repairs that can be dangerous to workers. Second, the intense heating step may damage the seals on storage canisters and cause them to leak. This releases hazardous wastes into the work area. Third, electrodes inside the glass-melting unit may also become damaged. When electrodes do not work correctly, an important aspect of vitrification is lost, that is, the controlled heating of wastes as they are mixed with the glass. Without properly functioning electrodes, technicians cannot monitor the heating step and so lose control over the reactions taking place inside the chamber.

The composition of the waste to be vitrified also affects the treatment's efficiency. For example, sulfate, molybdate, and phosphate compounds do not readily incorporate into glass. These elements when present in waste result in a heterogeneous glass block, which is less stable than properly vitrified material and perhaps less suitable for long-term storage. Chromium, aluminum, zirconium, iron, and bismuth exert a similar interference with the smooth heating and cooling characteristics of glass. Like most heterogeneous mixtures, glass that does not cool uniformly results in less long-term stability.

The Hanford and Savannah River sites led the way in improving vitrification to eliminate some of the disadvantages mentioned here. For instance, newer processes employ a stirring step when preparing the molten mixture to increase efficiency. Mixing distributes the wastes evenly and allows the glass to cool into a homogeneous solid. New glass formulas may also soon emerge with the capability to hold a larger proportion of wastes than the glass now used. Alternative formulas of aluminum-containing silicates, iron phosphates, or glass-ceramic mixtures hold promise in this area. Savannah River National Laboratory (SRNL) director G. Todd Wright said in a 2007 press release, "SRNL has a long tradition of putting science to work to create practical technology solutions that work."

A key way to make vitrification more efficient is to streamline the overall heating, mixing, and cooling steps. One innovation for accomplishing this streamlining is called the advanced vitrification system (AVS). AVS heats waste materials directly inside canisters, which after cooling become the permanent storage containers. AVS's advantage lies in the fact that treated wastes need never be removed from the reaction chamber. A process known as the Hohlraum melting method has further improved AVS. In the Hohlraum method, wastes and glass melt together as they are poured into the storage canister, increasing the treatment's efficiency. Next-generation vitrification methods will likely use conveyor belts of canisters that will move through the vitrification plant, filling up with vitrified wastes as they go.

The French nuclear industry has developed a new AVS concept that uses a cold crucible melter. In this method, a heating crucible (a type of heat-containing canister) raises the temperature of the wastes to an incredible 5,400°F (3,000°C), yet the entire canister's outer surface remains cool to the touch. Both standard AVS and cold crucible AVS treat twice as much waste as conventional vitrification, and they produce glass blocks half the size.

Most current cleanup/treatment technologies seek to keep the wastes at the cleanup site rather than transport them for treatment elsewhere. Vitrification with crucibles lends itself to this in situ treatment, meaning the activities take place right where the waste is located. In situ vitrification melts radioactive or nonradioactive materials in the ground where they are located. To perform this treatment, crews drive graphite electrodes into contaminated soil. The electrodes then heat the soil to at least

3,000°F (1,650°C) until it liquefies to a molten consistency. When the molten soil cools, its minerals coalesce into a glasslike material, which also reduces the size of the contaminated area by 30 to 50 percent. In the final step, bulldozers cover the treated area with clean soil.

AVS and in situ methods provide examples of ways to improve on vitrification's heating and cooling efficiency. As mentioned earlier, vitrification efficiency also depends in large part on separating the wastes into their constituents before treatment. Vitrification is like most other waste-treatment methods, especially thermal methods; efficiency improves when the composition of each waste load is homogeneous. Oak Ridge National Laboratory in Tennessee now develops "designer" solvents to selectively remove specific molecules from heterogeneous mixtures. For example, solvents called calixarenes remove cesium molecules from waste mixtures. Like calixarene, most solvent treatments target only one chemical at a time, so an obvious enhancement on the horizon will be the discovery of solvents that can remove more than one type of chemical from a single waste load. Perhaps this process will contribute to sustainability by making some of the extracted substances available for other uses. As early as 1995, Alex Gabbard of the Oak Ridge National Laboratory predicted that radioactive waste cleanup could contribute to sustainability. The waste site he had in mind was a contaminated pond full of radioactive sludge, and his team envisioned the use of crushed glass rather than solid glass blocks. Gabbard said, "The metals in the mud are a resource that could be sold for industrial use. If uranium can be recovered by this technique, it could be sold for energy production. After the metals are separated out, the remaining crushed glassy material could be reclassified as a waste material that requires low-cost disposal. This approach eliminates the need for long-term monitoring."

An apparatus developed in 1998 called an Archimedes filter (also called the Archimedes plasma mass filter) has now been adapted for today's vitrification technology for separating different radioisotopes before treatment. This may be especially helpful to the Hanford site, which holds nuclear wastes containing at least 99 percent heavy elements. The Archimedes filter vaporizes liquid wastes, then separates elements inside a cylinder using electromagnetic fields. Light elements (lighter than fluorine) are not ionized in these fields, and they gather at one end of the cylinder's chamber, while heavier elements drop out of the field to the chamber's inner surface.

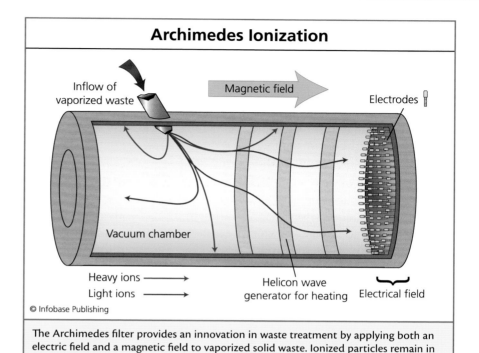

## Archimedes Ionization

Inflow of vaporized waste

Magnetic field

Electrodes

Vacuum chamber

Heavy ions →
Light ions →

Helicon wave generator for heating

Electrical field

© Infobase Publishing

The Archimedes filter provides an innovation in waste treatment by applying both an electric field and a magnetic field to vaporized solid waste. Ionized particles remain in the chamber and the clean vapor exits the filter.

Innovations such as the Archimedes filter present obstacles as does any new technology. In 2005, John Gilleland, director of San Diego's Archimedes Technology Group, admitted at a waste management conference that the new technology might not clean up all of Hanford's HLRW until the year 2028. He also added "it will cost billions of dollars and produce more glass logs than the proposed Nevada disposal site at Yucca Mountain has allocated to defense waste." Despite this gloomy opinion, or perhaps because of it, the waste management industry will likely try a number of technologies rather than rely on one.

Methods for making vitrification work easier, faster, and cheaper will help it become a standard method of hazardous waste treatment. Vitrification may someday replace incineration as the traditional method of nonhazardous waste treatment. To accomplish this, techniques for improving heating-cooling efficiency and waste constituent separation are necessary, and they must be able to work at reasonable costs.

# CONCLUSION

Vitrification immobilizes radioactive wastes in glass. In this form, solidi-fied waste can be safely stored or buried, and it is not likely to react with other chemicals in the environment and can last for a long time without corroding or leaching. Because of the chemical and physical features of glass, vitrified materials may someday become the safest of all treatment and disposal methods for the following four reasons. First, the structure of glass allows it to accept a wide range of chemicals, so it is suitable for radioactive wastes of unknown composition. Second, glass does not react with other chemicals and withstands wide temperature ranges and pressures. Third, glass is stable almost indefinitely. It is therefore a good matrix for holding radioactive chemicals with half-lives in the thousands to millions of years. Finally, molten glass can be molded into any shape before it cools, which helps in designing storage blocks, logs, or other shapes. Once it solidifies, glass will hold its shape for centuries.

Despite the promise of vitrification, it is not common in the United States. Only two nuclear reactor sites with very large radioactive waste vol-umes have begun to vitrify their hazardous stockpiles. Vitrification offers the advantages of being a large-volume treatment method and an in situ method. New technologies in heating-cooling and in crucible melting of wastes may make vitrification more efficient and faster. New types of glass are also being studied. These are intended to help the treatment work well on heterogeneous waste loads and may allow more waste to mix with smaller volumes of glass. By the time U.S. nuclear stockpiles have been reduced by half or so, these advances may be part of vitrification technology.

Though glass is inexpensive, vitrification itself is expensive because it uses large amounts of energy. Another disadvantage comes from the fact that some compounds in wastes are not compatible with glass. Glass has many attributes, but if cooled improperly it can solidify, crack, or give an otherwise poor final product. Vitrification technicians need intense train-ing in treatment techniques and radioactivity, and vitrification requires much advance planning and agreement from government and the local community on costs. Another drawback is that current vitrification does not reduce waste volume. Therefore the glass output must have a secure storage site large enough to accommodate the solidified blocks.

Vitrification's disadvantages are not insurmountable, and this tech-nology may offer the best hope for safe long-term storage of the world's most hazardous wastes.

# SOLIDIFICATION AND STABILIZATION

Hazardous waste does not always stay in one place. It travels from its source in shifting land sediments or in flowing waters. In worst cases, it enters an ecosystem and harms the health of plants and animals or kills them outright. That is why waste management emphasizes secure containment of excavated wastes or vitrified radioactive materials. The previous chapter showed how vitrification can turn an unwieldy volume of hazardous waste into a manageable solid. Solidification of liquids also helps keep wastes in one place and prevents their leaching into the environment. Creating a solid from a mixture of hazardous liquids can be thought of as a way to stabilize the hazards. This reduces the harm to the environment. Solidification and stabilization, therefore, work in similar ways to confine hazardous wastes to one spot that is easy to manage and monitor.

Waste-treatment activities such as excavation have disadvantages, notably the physical handling of tons of contaminated soil at a site, then the need to truck the loads to a separate location for cleanup/treatment. Leaks, spills, windy conditions, or accidents make hazardous-waste transport a risky business, and transport adds expense to each cleanup project. Neighborhoods may also resist the sight of waste loads rolling past their homes each day.

On-site incinerators serve well in treating wastes close to the spot where they are produced. Many manufacturing plants, factories, and hospitals use their own incinerators to treat daily waste loads. Incineration has, however, suffered from a series of mishaps and a bad image associated with air pollution. On-site vitrification fulfills many waste-management needs without a dark public image, but it is still years from being com-monplace. Chemical treatments such as oxidation and thermal wells work in situ, but these methods are not always available at a cleanup site. For each in situ treatment method, certain disadvantages can turn it into the wrong choice. Lastly, advanced technologies are sometimes too expensive for a budget-conscious project, such as found in *brownfield* cleanups.

Solidification and stabilization are gentle approaches to contamina-tion treatment. For example, a piece of land downstream from a mining site may be contaminated with acid runoff. In excavating the spoiled area, the cleanup crew might spill some chemicals, miss others, and probably spread the contamination. It is far less hazardous to the environment to simply neutralize the acids with bases—the hazard becomes stabilized and the wastes become safer for removal.

This chapter explores the techniques for controlling hazardous mate-rials in the environment without complex technology or tons of large equipment. It describes how solidification and stabilization offer simpler approaches to advanced treatment/cleanup methods, and it explores the consequences of uncontrolled waste streams in the sidebar "Case Study: The Sandoz Chemical Spill in Switzerland" found on page 113. This chap-ter covers the manner in which stabilization affects waste streams, the practical matters that must be considered in these types of treatments, and the new methods coming into use in solidification and stabilization.

## CONTROLLING SOLID AND LIQUID WASTE STREAMS

The concept of waste streams is relatively new in environmental sci-ence. Rachel Carson, discussed at length in the sidebar on page 106, led the way in explaining to the public the concept of a waste stream and its effect on living things. In her groundbreaking book *Silent Spring,* she described the problems of chemical wastes in the environment: "As crude a weapon as the cave man's club, the chemical barrage has been hurled against the fabric of life." Disposal of hazardous wastes has for too long meant dumping them into a river and letting the river carry the wastes out of sight. Of course other people upstream may well have been dumping their hazards into the same river. This indiscriminate disposal created a mixture of dangerous substances entering the ocean every day. Before most people became concerned about the environment, industries gave little care to chemical combinations or how they polluted eco-systems. In June 1969, a spark from a passing train ignited a fire on the Cuyahoga River in Cleveland, Ohio, and provided a clear sign of the risks of out-of-control discharge of chemicals into the environment. In addition to a bizarre scene in which bystanders watched a river burn, media coverage led the world to believe the water had ignited spontane-ously. Though the spontaneous ignition turned out to be false, the inci-dent became a symbol of an environment in crisis. A *Time* magazine article written after the fire described the situation: "Some river! Choco-late-brown, oily, bubbling with subsurface gases, it oozes rather than it flows. 'Anyone who falls into the Cuyahoga does not drown,' Cleveland's citizens joke grimly. 'He decays.' The Federal Water Pollution Control Administration dryly notes: 'The lower Cuyahoga has no visible signs of life, not even low forms such as leeches and sludge worms that usually thrive on wastes.' It is also—literally—a fire hazard."

Part of the new awareness about a damaged environment in the 1970s included ideas for arresting waste streams gone out of control. One way to control the movement of wastes is through stabilization by physical, biological, or chemical means. By stabilizing hazardous substances that have escaped from their normal waste stream, toxic chemicals may be pre-vented from entering and damaging ecosystems.

Waste streams can take many forms; they can be described in a broad sense or in detail. In the broad view of a waste stream, environmental sci-entists might consider the hauling of industrial waste by truck to barges that carry the loads to an incinerator. Waste streams may also include more local and detailed elements, such as a single city on the banks of one of the Great Lakes. The flow of liquid wastes from the city into the lake makes one waste stream with several components. These components may be any or all of the following: illegal or accidental waste spills plus runoff from farms and neighborhoods; oil leaks from a marina; ballast from a cargo ship; or solid wastes tumbling in from the shore. From above, a thunderstorm adds a deluge of vehicle exhaust chemicals caught in the atmosphere. The longer a person studies a single location, the more waste sources seem to materialize.

Hazardous waste migrates underground and is harder to measure and follow than surface waste streams. Underground waste streams develop silently, sometimes for decades, before an *aquifer* fills with a carcinogen or other deadly chemical. For this reason, the stabilization of hazardous chemicals includes a thorough understanding of geology and the ways in which the Earth moves.

## THE SEDIMENT CYCLE

Stabilization of underground pollution depends on the movement of land-masses on Earth. The characteristics of the land influence the rate of its shifts and movements. This is particularly important in waste burial, one of the most common methods of waste stabilization. Burial can be either in deep sediments under the land or beneath the ocean. Some in situ treatment methods leave the neutralized chemicals in the ground close to the surface. In situ vitrified wastes lie buried in a glasslike form in the upper layers of the Earth's crust, for example. The period of years in which these buried wastes remain stable and safe depends on the *sediment cycle,* also called the rock cycle.

The sediment cycle drives a slow movement of soils from molten rock under the Earth's mantle to the surface and around again. Unlike a nutrient cycle, a sediment cycle takes thousands of years to complete. It begins with soil. Soil is a loose mixture of inanimate (nonliving) bits and decaying organic matter covering the Earth's surface. Soil contains layers that each go through a gradual aging and migration process. The top layer that is high in organic matter is called the humus layer. Beneath the humus lies the topsoil, then the subsoil, and then bedrock. All the components from topsoil down to the Earth's mantle make up the sediment cycle.

Soil movements may be in an upward-downward direction, a side-to-side direction, or both. Soil on sloping land often undergoes an imperceptible downhill slide called *soil creep.* Soil creep imparts a sideways motion to the normal upward-downward movement of the sediment cycle, and it helps push contaminants toward the planet's surface waters.

Weather and heat provide the energy that powers the sediment cycle. Wind and rain wear on the Earth's surface and cause rock debris to erode. As rock erodes it moves toward low-lying areas, often under the sea. The deposited rock eventually forms submerged sediment that turns into sedimentary rock under constant and intense pressure from above. Meanwhile, heat combined with intense pressure in deep sediments also sets the sediment cycle in motion. These physical forces change sedimentary rock into metamorphic rock, which is the form that migrates upward as the surface erodes.

## The Sediment Cycle

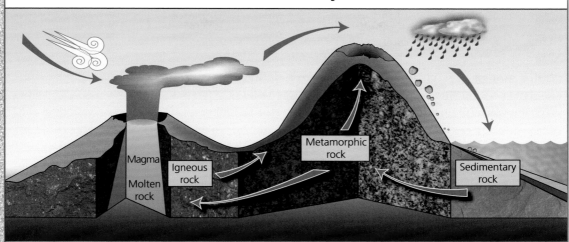

© Infobase Publishing

The sediment cycle consists of the world's most complex and slow-moving biogeochemical cycles. Though the cycle progresses over centuries, it works similar to other nutrient cycles by refreshing the nutrients available to plants and animals on the Earth's surface. Almost all elements connect in some way to the sediment cycle.

The Earth itself exerts two additional forces on the sediment cycle. One comes from the upward migration of igneous rock. Igneous rock begins as molten magma beneath the mantle, but upward pressure pushes it toward the surface, and it cools during this uplift. The other force comes through the movement of tectonic plates. As these plates collide, they thrust igneous rock and metamorphic rock upward and push sediments downward. A single sediment cycle would take millions of years without the power of tectonic collisions. Even so, cycles take thousands of years, except for one spectacular exception, volcanic eruption. Volcanoes force large amounts of materials upward all at once when molten igneous rock powers to the surface during a violent eruption. Volcanoes therefore help accelerate the Earth's rock and mineral cycles.

Though the sediment cycle normally takes thousands of generations, it cannot be ignored if permanent underground storage has been selected as a disposal method for hazardous waste. Buried radioactive wastes, for instance, must remain stable not for mere decades, but for many centuries.

## RACHEL CARSON

Few discussions on pollution and its effect on the Earth begin without mentioning the writings of Rachel Carson (1907–64). Born in the Allegheny Valley of Pennsylvania, she grew into a quiet and shy student devoted more to watching birds and studying nature than following the crowd. After finishing her undergraduate work, she earned her Master of Science degree in zoology from Johns Hopkins University. Carson began her career as an aquatic biologist who jotted observations as she accompanied research teams to the ocean. She began publishing on the relationships between life on land and in the sea in 1941. Her breakthrough work, *The Sea Around Us*, published in 1951, described how human activities affected the health of the oceans and marine creatures. But 1962's *Silent Spring* made a greater impact by drawing attention to the tremendous amount of chemicals entering ecosystems every day. In it she examined the routes of pesticides into food chains. *Silent Spring*'s focus on dichlorodiphenyltrichloroethane (DDT) became a milestone in environmental awareness. Carson traced the progression of DDT through the environment to demonstrate how waste streams threaten species dependent on each other for life. Said Carson about her writings on the environment, "The more I learned about the use of pesticides, the more appalled I became. I realized that here was material for a book. What I discovered was that everything which meant most to me as a naturalist was being threatened, and that nothing I could do would be more important."

Carson attracted little interest from the academic community through most of her career, including her warnings about pesticides. Undaunted, she conducted her own studies. Carson focused on the toxic consequences of spraying DDT onto fruit trees, which then leached from orchards into ecosystems dependent on insects, songbirds, and raptors. *Silent Spring* became the first work to shed light on the near extinction of bald eagles and peregrine falcons due to the damage done on their reproduction by DDT. Though the term was not yet coined, *Silent Spring*'s description of DDT poisoning in eagles introduced the idea of bioaccumulation.

Rachel Carson endured harsh criticism from the pesticide industry and other groups during her writing career. Fellow scientists dismissed her as an irrational woman who was hysterical over nature, not unlike the criticism directed at today's tree huggers. Edwin Diamond wrote in the *Saturday Evening Post* in 1963, "Thanks to an emotional, alarmist book called *Silent Spring*,

## SOLIDIFICATION

Solidification, also called immobilization, does not destroy hazardous wastes but instead holds these substances in a form that will not move or leak. Solidification usually involves mixing wastes into a stable

Americans mistakenly believe their world is being poisoned." Yet environmental studies completed in the years following her death agreed in general with Carson's theories and research. Her writings inspired the young environmental movement and led to the banning of DDT in 1972. In *Silent Spring*'s 40th-anniversary edition, Carson biographer Linda Lear described the essence of Rachel Carson's place in the environmental movement. "Headlines in the *New York Times* in 1962 captured the national sentiment: '*Silent Spring* is now noisy summer.' In the few months between the *New Yorker*'s serialization of *Silent Spring* in June and its publication in book form that September, Rachel Carson's alarm touched off a national debate on the use of chemical pesticides, the responsibility of science, and the limits of technological progress. When Carson died barely eighteen months later in the spring of 1964, at the age of fifty-six, she had set in motion a course of events that would result in a ban on the domestic production of DDT and the creation of a grass-roots movement demanding protection of the environment through state and federal regulation." Today's scientific scrutiny of pesticides and other chemicals' effects on human, animal, and aquatic health originated with the vision of Rachel Carson.

"For the first time in the history of the world, every human being is now subjected to contact with dangerous chemicals, from the moment of conception until death. In less than two decades of their use, the synthetic pesticides have been so thoroughly distributed throughout the animate and inanimate world that they occur virtually everywhere." —Rachel Carson, *Silent Spring* (NOAA/U.S. Department of Commerce)

material called a solidifying agent that does not react with substances around it.

Solidification technologies handle from small amounts to very large volumes of contamination. Small volumes of solidified wastes are usually stored in 55-gallon (208-l) drums and taken to a landfill. Larger volumes require mixing with the solidifying agent in a mobile machine called a pugmill. A waste treatment firm may own one or more pugmills, which workers tow onto a hazardous waste site and set up near the contamination. After mixing and solidifying, they either bury the solids in the originally polluted area or haul it away to another site for burial or landfilling. Very large volumes of contamination may be too great for even the biggest pugmills, so these are treated by in situ solidification where workers add the mixing agent directly into the contaminated soil.

In situ solidification suits the nation's worst Superfund sites because it reduces the need to transport toxic materials until they have been converted to a safer form. Solidification methods can be adjusted to accommodate various volumes of contamination and varying depths in soil. The two most common solidification methods are in situ treatment and deep-soil treatment. In situ solidification works on tons of excavated contaminated soils that must be mixed with the solidifying agent before being reburied. A disadvantage of this method is that some contamination may be missed by the excavation and left in the ground. Deep-soil solidification avoids the problem of partial excavation because in this method powerful mixers burrow into contaminated soil to depths of 150 feet (45.7 m) and then dispense the solidifying agent. As a result, the solidified waste-soil mixture stays where it was treated. Very large deep-soil mixers involve mechanical cranes that support a drive shaft and long mixer equipped with paddles and augers. The mixer digs through the soil as the shaft moves downward, and then the operator pumps solidifying agent into the drilled well. After soil, waste, and solidifying agent have reacted, the crane pulls the solid cylinder out of the ground. The crane therefore extracts a solid column of immobilized waste of 35–40 inches (89–102 cm) in diameter. Trucks simply transport the column to a final disposal site.

Solidification obviously depends on good quality solidifying agents. Some characteristics of good solidifying agents for wastes to be treated and left in the ground are the following: (1) resistance to degradation in soils saturated with toxic chemicals; (2) compatibility with the chemicals in wastes; (3) non-permeability to groundwaters; (4) ability to bear heavy weight without crumbling; and (5) being nonreactive with water or soil.

Cement meets the requirements listed here and is a common solidifying agent used in the waste industry. Cement makes a good solidifying agent because it is inexpensive and mixable. Several pollution cleanup companies now make specially formulated cements tailored to the type of contaminants and the soil's properties at hazardous sites. Cement formulation starts with the preparation of slurries (semisolid mixtures) that contain different moisture levels. These slurries usually consist of clays called bentonites or attapulgites. Excess fly ash recovered from incinerators also adds consistency to solidification formulas. Finally, softening agents in the formula help the blending of waste, soil, and cement components.

Many brownfield sites and about one-quarter of Superfund sites now use what the Environmental Protection Agency (EPA) terms *S/S* for solidification/stabilization. The EPA identified S/S in its 1999 "Solidification/Stabilization Resource Guide" as one of the top choices for Superfund source control treatment technologies. Source control methods are technologies designed to treat wastes at the place where they have originated. S/S works well on lead, arsenic, petroleum, and creosote contamination because these agents react readily with solidifying agents. The sediments below New York Harbor contain all of the above chemicals plus other heavy metals, dioxins, and polychlorinated biphenyls (PCBs). S/S makes a good treatment choice in this location because the contamination rests under a busy waterway.

S/S currently makes up about 6 percent of in situ methods and 18 percent of all aboveground methods in hazardous waste site remediation. Hospital wastes may also soon be treated by S/S to reduce the total amount of waste that goes to the hospitals' incinerators. Solidification is particularly useful for hospital wastes because it makes sharps such as needles, razors, glass, and dental devices safer to handle and transport.

## CHEMICAL AND BIOLOGICAL STABILIZATION

Stabilization can be done by either chemical means or biological reactions. Solidification of wastes with materials such as cement also represents a type of chemical stabilization. Solidified waste is more stable in the environment than liquid or poorly contained solid wastes.

Chemical S/S methods rely on materials other than cement in certain circumstances. For example, the chemical lime (calcium oxide, $CaO$) in combination with incinerator ash extracts many toxic metals from soil. The metals react with lime in the presence of water to form metal hydroxides, which turn into a cementlike solid in the presence of ash and water:

$$lime + fly\ ash + waste + water = final\ solid$$

In this reaction, waste makes up 8 to 20 percent of the solid's final weight; water accounts for more than 50 percent. Organic compounds undergo a similar solidification reaction but make up only 10 to 20 percent of the final product's weight.

Pozzolan offers an alternative to cement and is used in a manner similar to lime-fly ash. Pozzolan is a heterogeneous powder from the Earth's crust composed of diatomaceous earth, volcanic ash, shale, and pumice. Pozzolan substitutes for fly ash in the S/S equation and reacts with hydroxides to form a cementlike substance, which can immobilize either hazardous metals or nonmetal compounds.

*Calcination,* also called calcining, is a recent development in solidification in which heat solidifies wastes. Calcination involves three steps: (1) chemical conversion of the waste; (2) reaction with the minerals in soil; and (3) heating to create the final solid form. Calcination may prove to be most useful for treating liquid high-level radioactive waste (HLRW) so that the wastes are rendered safe for transport to a disposal site.

Some treatments, such as precipitation (also called agglomeration), combine physical and chemical methods. Precipitation is a good choice for cleaning up sites contaminated with toxic metals attached to soil. This is because metals tend to attach to moist soils, but they detach from the soil and form a precipitate in the presence of hydroxide chemicals; detachment works best in the presence of sodium hydroxide (caustic soda) or calcium hydroxide (lime slurry). The reaction forms a small chunk of metal hydroxide called a floc. Finally, asphalt, cement, or pozzolan can be used to immerse the flocs and thus hold them in a stable, safe form.

Encapsulation takes place by immersing waste particles in an inert material such as asphalt. In this way encapsulation turns nuclear wastes and other hazardous wastes into a stable solid, which is physically unable to further contaminate the environment. Other industries besides waste treatment—medicine, cosmetics, food production—also use encapsulation as a delivery system for products or preservatives.

Thermoplastic microencapsulation offers an innovation in encapsulation in which heated material encloses each individual floc. Microencapsulation begins by drying nuclear, metal, or industrial wastes and then dispersing them in melted asphalt or plastic. A machine forces the mixture through a sieve to turn it into tiny granules and, after the granules cool, the waste mixture has been transformed to a coated end product. Encapsulation requires specialized equipment to heat the mixtures and form granules that will hold together. The process resembles the steps used to make dough because there is no set recipe for getting a desired consistency. Small businesses can handle this variability in the method, but that same variability would cause problems in large-scale treatments. Until these details have been resolved, encapsulation may play only a small role in stabilization technology.

Biological stabilization called *phytostabilization* uses vegetation rather than chemicals to keep hazardous substances from moving through the soil. Phytostabilization provides an advantage over chemical methods and encapsulation because it relies on natural forces; it requires only a tree, grass, or plants with a good root system to hold contaminants in place.

Plant life immobilizes organic and inorganic chemicals in three main ways. First, plant roots pull water out of the soil and, by doing so, keep soluble wastes from leaching into groundwaters. Plants that transpire large amounts of water are best at stabilizing wastes because they draw large amounts of water from the earth. Second, plant roots adsorb molecules such as heavy metals. Adsorption is the process of capturing a molecule by adhesion to a surface. The adsorbed metal accumulates on the outside of roots and travels no farther through the soil. Third, roots excrete chemicals or enzymes that precipitate soluble compounds. The precipitated particles then adhere to soil through electrostatic attractions instead of leaching away.

Phytostabilization has been tried in hazardous waste cleanup/treatment in the United States. Superfund sites in Bunker Hill, Idaho, Palmerton, Pennsylvania, and Whitewood Creek, South Dakota, each use vegetation to stabilize toxic mining wastes and metals. The U.S. Army's Aberdeen Proving Ground in Maryland has planted more than 200 poplar trees to remove a plume of organic solvents contaminating the base's aquifer. New York's Brookhaven National Laboratory Technology Fact Sheet on these sites summarizes the value of phytostabilization: "Phytostabilization is considered less disruptive than excavation. In addition, its compatibility with the wetland environment and aesthetic factors make it a desirable choice."

Starting in 1947 and for several years following, Brookhaven National Laboratory contaminated its land with PCBs, heavy metals, and radioactive chemicals. An aquifer became spoiled, the compounds leached into deep sediments, and the nearby Peconic River took in the pollutants. In 1989, the EPA identified Brookhaven as a priority cleanup site but the contamination persisted for many years afterward. Environmentalist Scott Cullen said in 2000 in the *New York Times*, "We know that there are concentrations of cesium-137 and strontium-90 in Peconic River fish at two to three times the levels of other areas." More than $353 million has been invested so far to excavate 55,000 cubic yards (42,050 m³) of contaminated soil and to extract 5,000 pounds (2,268 kg) of contaminants from the aquifer. Meanwhile the Peconic's restored wetlands have prevented pollutants from entering the river at zero cost and without need for manual labor. Few cleanup/treatment methods used today are 100 percent effective, so small amounts of PCBs and toxic metals still spoil low-lying areas near the Peconic River, but the work continues for removing these amounts as well. In 1999, the U.S. Department of Energy

In Krefeld, Germany, a church and surrounding residential area lie between power and chemical plants and the Rhine River. Since the Sandoz chemical spill in 1986, countries along the Rhine have made progress in cleaning out metals and chemicals, but fertilizer runoff continues to spoil stretches of the river. *(Michael Utech)*

## CASE STUDY: THE SANDOZ CHEMICAL SPILL IN SWITZERLAND

On November 1, 1986, the BBC reported, "There has been a catastrophic fire at a chemicals factory near Basel, Switzerland, sending tons of toxic chemicals into the nearby river Rhine and turning it red. The fire broke out early this morning in a storage building used for pesticides, mercury and other highly poisonous agricultural chemicals. People in Basel and the surrounding region on the border between Germany and France were told to stay indoors. Witnesses reported a foul smell of rotten eggs and burning rubber. Fourteen people, including one of the firemen fighting the blaze, were treated in hospital after inhaling the fumes." The warehouse site belonged to Sandoz A.G. chemical company from which 30 tons (27 metric tons) of contaminated water flowed past banks and infiltrated river ecosystems. Environmentalists estimated at least 500,000 fish died because of the spill, and it affected wildlife and plant life 120 miles (193 km) from the warehouse. Physician Thomas W. Clarkson of the University of Rochester Medical School was quoted by the *New York Times* regarding his opinion of the health threat: "It's clearly being diluted if it's moving downriver. The big danger would be if the mercury stays in one place. If that's the case, the conditions could in fact convert it to methyl mercury and then it would get into the flesh of the fish. And that's the danger, if people consume the fish." The spill polluted parts of Germany, France, and the Netherlands, in addition to Switzerland as the waters headed toward the North Sea. The BBC noted that in a single day the Sandoz spill had reversed 10 years of cleanups that had been taking place along the industrialized Rhine.

The Sandoz spill demonstrated the ability of waterborne chemicals to travel great distances. Though no one determined the spill's full effect on river and marine biota, the accident compelled local governments to more closely monitor the river all the way to the sea. Investigators hired by local municipalities to study the river's contamination soon discovered that companies other than Sandoz had been deliberately dumping hazardous wastes into the Rhine for years. As a result European environmental laws strengthened, and environmental organizations on the continent gained more influence with their governments.

*(continues)*

*(continued)*

A new field of study called *ecotoxicology* developed as a consequence of the Rhine investigations. Ecotoxicology is the study of pollution's toxic effects on all the biota in an ecosystem. Studies are set up in one of two ways: (1) collecting data from wildlife, plants, fish, invertebrates, and microbes in nature or (2) using laboratory models. Data collection from wildlife can be difficult and time-consuming because of the efforts needed to find sick animals and take samples from them. Many biological systems consisting of invertebrates, fungi, or small plants may never be sampled in an ecotoxicology field study, adding uncertainty to the entire study. Laboratory models avoid these disadvantages because scientists study the effects of toxic compounds directly on mammalian tissue. Studies on living tissue conducted in laboratories rather than in nature are called in vitro tissue culture. In vitro experiments now contribute much of the data gathered in ecotoxicology studies, such as those conducted after the Sandoz accident.

Today the EPA's EcoTox program provides information on the effects of hazardous chemicals on terrestrial and aquatic life. The Sandoz accident taught scientists and the public about the far-reaching effects of pollution. It also demonstrated that environmental damage rarely confines itself to just one small and easily managed area and can be felt for many years.

(DOE) expressed skepticism about the use of plants to clean up the river, a general process called *phytoremediation.* (Phytostabilization is a type of phytoremediation.) A DOE spokesperson told the Riverhead, New York, *News Review,* "The effectiveness of phytoremediation in reducing the contaminant concentrations to the cleanup objectives is uncertain." Since then, a modest number of cleanup projects have shown phytoremediation supplements other cleanup methods. A 2007 status report published by the EPA titled *Treatment Technologies for Site Cleanup* included phytoremediation as one promising new technology for groundwater and wetland cleanup, such as those along the contaminated portions of the Peconic River.

# UNDERGROUND DISPOSAL

Radioactive wastes must be stored until they have decayed to safe levels, an event that may take thousands of years. Today, every place that holds a radioactive stockpile takes into account three factors for meeting the demands of long-term storage: time, distance, and shielding. The time people are exposed to high doses of radiation should be minimized. An ideal exposure time is zero minutes, but of course this does not occur in the nuclear industry. The distance between a radioactive stockpile and nearby homes and workplaces should be as great as possible. Finally, protective shielding should provide adequate radiation-absorbing barriers to reduce to zero the emissions that hit anyone's body. Burial, also called geological burial, has been accepted as the best method for attaining these three safety measures.

Sir David Wallace of the United Kingdom's Royal Society, an organization of leaders from various sciences, explained in 2006 to London's *Independent* newspaper, "The nature of scientific knowledge is such that there will always be levels of uncertainty associated with any method of disposing of radioactive waste. However, there is considerably less uncertainty surrounding burying radioactive waste deep underground in stable geological formations than other options." Waste burial in deep sediments is also known as deep-well injection, deep-well disposal, or subsurface injection. This method of waste stabilization has been used by the United States for more than 30 years with few mishaps, and it is inexpensive relative to thermal treatments. The major concern about deep burial arises from potential instabilities of the land. Some of these seismic threats to underground waste storage are described in the sidebar "Yucca Mountain Disposal Site" that follows on page 118.

*Subduction* waste burial has been proposed as a safer alternative to simple burial, but it has the disadvantage of being very difficult to accomplish. In subduction waste burial, crews would bury wastes within stable clays in a tectonic plate that will slowly move under an overriding plate (a subductic plate) and carry the waste deeper into the Earth's mantle. Several subduction plates are known to meet thousands of feet below the ocean's surface, plates that could possibly serve to move wastes far from humans. Geologists and nonscientists have long argued the benefits and risks of this method of waste disposal. In 2005 *New Scientist* magazine described the challenges to overcome if this disposal method were ever attempted: "Subduction zone insertion is perfectly sound in theory, but there are significant practical problems. The zones are inherently unstable and unpredictable, and the sediment on top of the subducting crust plate tends to get scraped off rather than being carried into the mantle . . . This could lead to waste being squeezed back to the seabed in the future. Drilling it deep into the basalt of the crust may solve this, but at the depths typically encountered in subduction zones, drilling is all but impossible."

Submersion in water may offer an alternative to deep burial and be more practical than subduction. Rather than the difficult task of burying wastes under the ocean floor, waste loads stored in secure containers enter constructed pools or lagoons. Proper submersion can meet requirements for time, distance, and shielding, but it is difficult to monitor if a leak or corrosion should occur. Vitrified wastes might lend themselves to this approach because the glass form of waste does not leach into water.

Aboveground dry storage casks offer a less expensive storage than underwater methods and are easier to maintain and monitor. As long as they fulfill storage requirements for time, distance, and shielding, dry casks may be a feasible approach to HLRW storage. The aboveground casks may, however, remind nearby communities of the radioactive wastes constantly within their sight. Visible aboveground casks might also serve as targets in the rare event of a terrorist attack. Both aboveground and underground casks possess another disadvantage: potential damage from earthquakes.

Deep burial is done by pumping liquid wastes through a pipe drilled vertically into the earth until it has reached porous layers far below any aquifers. A deep well drilled at the Rocky Mountain Arsenal near Denver in 1961, for example, extended 12,000 feet (3,658 m) or 2.27 miles (3.66 km). Wastes flow to the well bottom and exit the pipe and then porous rock absorbs the hazardous materials. Hard impermeable rock usually surrounds the more spongelike porous rock, so deep burial should prevent chemicals from contaminating underground water sources, but no one knows for sure what happens at that depth. In a 2007 EPA press release, the agency explained why it approved the use of deep-well disposal of chemical wastes made by the Occidental Chemical Corporation in Wichita, Kansas: "EPA concurs with Occidental that this method of disposal is protective of human health and the environment and is cost-effective. The method isolates the waste in deep geologic formations with no potential for future contact with underground sources of drinking water or the environment above the surface. Based on a technical review of Occidental's petition,

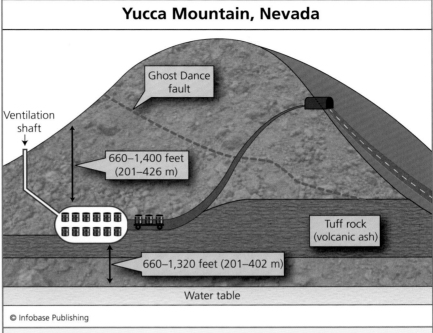

**Yucca Mountain, Nevada**

Ghost Dance fault

Ventilation shaft

660–1,400 feet (201–426 m)

Tuff rock (volcanic ash)

660–1,320 feet (201–402 m)

Water table

© Infobase Publishing

When the Yucca Mountain nuclear repository begins operations, the U.S. Department of Energy plans on storing 77,000 tons (70,000 metric tons) of nuclear waste there. The department plans to send the first shipment to the repository in 2017.

EPA found that the company satisfactorily demonstrated that the hazardous waste will not move out of the injection zone, the base of which is nearly one mile [1.6 km] deep, for 10,000 years." Occidental's study was based on accepted mathematical models for predicting the movements of chemicals through various types of rock and sediments.

Despite the EPA's optimism, waste experts such as environmental engineer William Blackman of Arizona State University point out a list of concerns regarding deep-well disposal, as follows:

- expensive construction and operation
- sudden changes in injection pressure, possibly leading to leaks
- constant monitoring to show no contamination of aquifers is taking place

- disintegration of casing or damage from seismic activity may cause leaking
- requires large open spaces in the rock formation to prevent displacement of clean aquifers

## YUCCA MOUNTAIN DISPOSAL SITE

In 1985, the DOE announced plans to use a site located 100 miles (161 km) northwest of Las Vegas and 25 miles (40 km) from California's Death Valley for underground disposal of the nation's HLRW. The site is named Yucca Mountain. Though deep burial of HLRW was perceived as the safest disposal method for these environmental hazards, a public debate soon ensued over whether Yucca Mountain offered the best features for such a grave task.

Arguments center around two factors: (1) the safe permanent storage of the materials deposited at the site and (2) the transport of wastes to the site. Opponents questioned the safety of the site in case of earthquake damage, volcanic activity in the area, or attack by terrorists. The DOE launched several studies to address these concerns, and after one risk assessment report had been made public, undersecretary of the DOE in 1998, Ernest Moniz, said, "Obviously, in our view, it is a very serious document. Everyone, in my view, is certainly impressed with the amount and quality of the work that has to be done." Senator Richard Bryan countered in the same *Las Vegas Review-Journal* article, "Surprise, surprise. They like their own work. That is all this document represents. After twenty years of study, major questions of the suitability continue to linger and, indeed, have been heightened."

These harsh words marked the tip of the iceberg in the Yucca Mountain debate. Indeed, the Yucca Mountain structure may be laden with tiny fissures; at least 32 active earthquake faults run through the region, and the area contains an active volcano. Scientists and nonscientists fear that liquid wastes will leak into rock fractures and cause a massive explosion. The state of Nevada pitted itself against the DOE's construction plans. Yucca Mountain soon became a symbol of the nuclear industry's possible flaws and the public's alarm over radioactive materials in their surroundings. In 1998, Nevada governor Bob Miller wrote to Secretary of Energy Bill Richardson, expressing his constituents' concerns: "The Yucca Mountain site is not suitable for development as a high-level nuclear waste repository. It should be immediately removed from consideration for a repository because it meets the conditions of the Department's [DOE] guidelines for disqualification with respect to the rapid flow of groundwater from the proposed repository to the adjacent environment." The governor also expressed unease over

Another weighty challenge of deep burial resides in the selection of a suitable site for drilling a well. The seismic activity of the area, that is, its history of earthquakes and volcanoes, must be thoroughly studied before any drilling begins. Any site having a medium or high probability

seismic instability and volcanic activity within the mountain. The Yucca Mountain project nevertheless continued. In the past 10 years most Nevadans have opposed both the storage site and the rail shipments of nuclear waste through their state to the repository. In 2006, Nevada senator Harry Reid summed up the feelings of the repository's staunchest opponents by stating in the *Las Vegas Review-Journal*, "It doesn't matter where the railroad goes because Yucca Mountain will never happen."

The DOE has balanced its need to placate worried residents along the shipment routes with a critical need to put the nation's nuclear stockpile somewhere that is a long-term safe solution. Spent nuclear fuel and HLRW from nuclear power facilities and national defense programs now sit at 126 separate sites around the country, as much as 14 million cubic feet (400,000 m$^3$) in total. Some of the buried drums have been stored for so long they are corroding. Commercial nuclear power plants also have their own stockpiles of spent fuel rods, usually submerged in storage pools on each facility's property.

The project has been plagued by errors in planning and miscalculations in safety assessments, so delays have extended the date on which Yucca Mountain will accept its first waste load. The enormous expense of building this large underground repository has created the most difficult obstacle to the project. Ward Sproat, director of the Office of Civilian Radioactive Waste Management, told the *Las Vegas Review-Journal* in 2008, "Until we get this issue (budgeting) fixed I can't, nor can anyone else, tell you with any degree of certainty when the repository is going to open. This is the single biggest issue we as a country need to address so this repository can go forward." As of 2008, Mr. Sproat estimated that the site would not be completed before 2017, and he projected the cost to exceed $77 billion.

Aside from costs, what information exists on Yucca Mountain's safety? No one knows, but the EPA stated in 2008 that it had adequately showed in experiments that the storage would be secure. EPA administrator Jeffrey Holmstead assured the public, "EPA met this challenge by using the best available scientific approaches and has issued a [public health] standard that will protect public health for a million years."

of earthquakes should be rejected for deep burial and only areas far from any volcanic activity should be considered. Soil and deep sediment conditions (moisture, pH, and mineral components) also affect the materials that make up the well's outer casing and so these things must be compatible with the well casing. Underground factors are tricky to monitor and sometimes difficult to predict, so many environmental scientists feel that deep-well burial places too great a risk on the environment. Alan Farago, chairman of the Miami, Florida, chapter of the Sierra Club, has opined, "The premise of such injection is that [treated wastewater] burial, thousands of feet underground, is the last we will ever see of it. It is an easy thing to believe because we wish nothing more than to be far removed from our waste, but it is not true. A toxic stew of ammonia, fecal coliform, and volatile organic chemicals is rising to meet our drinking water through the very underground injection control wells put in place to rid us of the waste."

One peculiar event at the Rocky Mountain Arsenal demonstrated another potential hazard from deep-well drilling that may give many people pause. During Rocky Mountain's cleanup in the early 1960s, crews built a large-capacity deep well, 12,045 feet (3,671 m) in depth, for storing hazardous waste. Over a four-year period starting in 1961, the U.S. Army injected millions of gallons of nerve gas–contaminated water into the shaft. During those same years, the nearby Denver area experienced at least 700 mild earth tremors, which seismologists found surprising since Denver had not had an earthquake in 80 years. The U.S. Geological Survey's (USGS) Web site describes the incidents: "In 1961 a 12,000-foot [3,658-m] well was drilled at the Rocky Mountain Arsenal, northeast of Denver, for disposing of waste fluids from arsenal operations. Injection was commenced March 1962, and an unusual series of earthquakes erupted in the area shortly after." Debates began among the public, government, cleanup experts, and scientists as to the cause of the tremors. Many people believed the drilling into deep sediments had disturbed pressures deep below the arsenal. By 1966 *Time* magazine reported, "Then consulting geologist David Evans suggested that the quakes under the suddenly shaky Colorado terrain could be traced to a deep well at the nearby Rocky Mountain Arsenal. Military and civilian experts scoffed, but Evans backed up his theory with impressive evidence." Mr. Evans proposed that the injected gallons had "lubricated" the surfaces of fractures within deep rock formations, allowing them to more easily slide past each other, thus causing tremors. Evans's studies did not prove that the well caused the tremors, but it did show that the epicenter of activity occurred only a mile from the well and at 12,000 feet (3,658 m). The USGS at the time admitted that the Colorado earthquakes and the army's underground disposal system "probably are related." The U.S. Army closed the Arsenal's well as a precaution in 1966 and permanently sealed it in 1985.

The United States currently has five deep-well facilities run by commercial companies, two in Texas and one each in Louisiana, Oklahoma, and Ohio. These sites accept wastewaters containing heavy metals, organic compounds, nonbiodegradable materials, and cooling waters from manufacturing plants. They require that the waste liquids are first filtered to contain no more than 100 ppm (parts per million) of suspended solids and be nonacidic, or pH 6.0 to 12.0. These features help prevent corrosion and help make the pumping process (injection) easier. The federal government and some municipalities operate additional deep wells throughout the United States, but there exists little public information on the details of these wells.

## NEW TECHNOLOGIES IN WASTE STABILIZATION

New technologies in waste stabilization focus on safety and durability. If chemicals are used as stabilizing agents, they must remain stable in the environment at least as long as the wastes persist. Certain plastics have characteristics that may fulfill these needs, and the cement industry continues working on products specially formulated for various wastes.

The plastics industry offers two different technologies for waste stabilization. The first technology uses polymers, which are very large, long compounds made of repeating subunits. For example, starch is a polymer of the sugar glucose. The second technology called thermoplastics involves reactive polymers, which are chains that form when a mixture of subunits is heated. The heat provides the energy needed for binding one subunit to the next. After the polymer forms and cools, it sets into an inert rigid material that is stable when exposed to heat or cold. In other words the polymer has all the characteristics of a good stabilizer.

Polymer chemists study different types of subunits to build new polymers that will improve waste stabilization. Two types of polymers have promising attributes: resins and polystyrene. Resins form in alkaline (pH greater than 7) conditions by adding sodium hydroxide to a mixture of subunits. Polystyrene is a chain of styrene molecules that bind in the presence of a chemical called peroxide. Both of these synthetic polymers offer a key advantage as stabilizing agents: synthetic compounds are not found in nature and are therefore not easily degraded. (They are nonbiodegradable.) Polymers made by plants and microbes would not work well in waste stabilization because the natural compounds degrade in soil.

In thermoplastic technology, wastes mix into compounds that have been melted, and, when the mixture cools, a rigid, solid plastic surrounds the waste (similar to encapsulation discussed earlier). Examples of thermoplastics for waste stabilization are the polymers polyethylene and polypropylene. Another material that works well is asphalt. The asphalt industry makes its product from the gooey material left over when refineries distill crude oil to make fuels and lubricants. Bitumen is a similar product made from coal. (Countries other than the United States use the term *bitumen* for asphalt.) Asphalt or bitumen offers an inexpensive approach for holding wastes in place in the environment.

Cement technology focuses on creating new formulas and additives for adjusting the properties of the cement-waste mixture. Gypsum, lime, and portland cements hold large amounts of waste and are durable and hard. Portland cements make up a special type known as hydraulic cements; they use water for hardening. Once they become solid, portland cements are impervious to water, and so they may be the best choice for stabilizing wastes in moist soils.

Polymers and cements work best in underground stabilization of wastes, while plants work on chemicals closer to the surface. For instance, plants making up a general group called ground covering carry out phytostabilization by holding contaminated topsoil in place. Specialized aspects of phytostabilization include plants that extract contaminants from the earth rather than merely stabilizing them. Plants give an added advantage because they help rebuild ecosystems, but cements and polymers do not.

Stabilization under water is more difficult than stabilization on land. Cleanup crews have a difficult time reaching wastes in sediments under rivers and harbors, and they also face the challenge of keeping hazardous materials from escaping with the currents. Many underwater waste deposits furthermore contain high levels of PCBs and pesticides, which aquatic plants are not effective in stabilizing relative to other chemicals. Both chemical and biological technologies therefore have room for improvement for stabilizing underwater contaminants.

# CONCLUSION

Solidification is a way to keep wastes from moving through soils or waters. In this method, a chemical solidifying agent added to a contaminated area captures hazardous chemicals and holds them in a stable solid form. The formation may either stay at the originally contaminated site or be transported to another site for disposal. In situ solidification uses the soil itself as the matrix in which wastes are solidified so it is valuable for large contamination sites.

Solidification stabilizes waste, which means the waste is made stationary so it cannot contaminate the surroundings. In addition to solidification, waste may be stabilized by burying it, by chemically treating it, or by treating it with living things. Deep burial can be thought of as a way to stabilize waste, because it puts the waste in a secure place far from animal and plant life. The sediment cycle affects the safety of deep burial wells, so this method requires thorough knowledge of the land's geography. One proposed burial site for radioactive wastes has been the Yucca Mountain Disposal Site in a remote part of Nevada. This site has been plagued by controversy since its beginning and has not yet accepted any HLRW. Many of the problems experienced at Yucca Mountain may represent potential hazards at other underground waste storage sites, that is, potential leaking into groundwaters or damage by shifting landmasses. While the United States seeks to find a permanent waste storage site that everyone can agree on, radioactive wastes accumulate in temporary storage at nuclear plants around the country.

Chemical stabilization involves materials that envelop small particles of waste. *Encapsulation* is the term used for this process. In microencapsulation, the material envelops each individual particle of waste called a floc. Asphalt and new synthetic plastics have led the way in encapsulation technology. Another option for stabilizing hazardous waste is by the use of plant life, a process called phytostabilization. Plants or trees with extensive root systems that absorb large amounts of water draw hazardous materials out of soils. This method is not disruptive to the environment and helps restore ecosystems as the plants grow.

Both solidification and stabilization avoid the use of heavy machinery, trucks, and complex equipment. New techniques are being studied to target specific types of waste and to treat larger waste loads. Solidification and stabilization are not highly technical methods, and, other than deep burial, they have proved to be environmentally safe and therefore good choices for hazardous waste treatment.

# REDUCTION AND COMPACTION

Waste creates a problem in the environment because it takes up space. No one wants to live near a trash dump, and certainly no person would choose to build a home next to a hazardous waste site. To keep waste out of sight, landfills and heavy industry for many years were situated far outside town limits. But as populations expand outward from urban areas, more and more houses occupy land once used for low-density activities; an area called the urban fringe. When urban areas spread, space for municipal solid waste (MSW) or more hazardous types of waste becomes scarce. Minimizing total waste volume therefore helps manage the waste problem. Reduction is any decrease in the amount of waste by physical methods or other means; compaction is the physical decrease in the volume of waste. *Volume reduction* is another term for compaction.

The best way to reduce the volume of nonhazardous waste comes from source reduction. In source reduction, people and businesses avoid generating waste in the first place. Manufacturers do it by making products and packaging that contain less unusable materials. Sustainable manufacturing, which is becoming more and more common in European countries, depends on new product designs to reduce scrap materials and use less raw materials. This type of source reduction is not yet commonplace throughout the world, however, so nonhazardous wastes continue to mount.

The remarkable growth of consumerism in the past three decades has complicated the task of reducing the world's volume of waste produced every day. Since 1970 the gross world product, which is the sum of all goods and services produced globally, has more than tripled. Who is using all these products? The 359,000 people born each day (250 newborns every minute) certainly contribute, but population increase only partially explains the rise in consumption. The bulk of the world's consumption takes place in industrialized countries, while developing countries own the greatest increase in consumption rate. China and India provide examples of rapidly growing economies linked to an increase in consumerism. In many parts of the world, economic growth may play a more vital role in people's lives than resource conservation. The Worldwatch Institute has calculated that the planet has 4.7 acres (0.02 km$^2$) of land available for each person's resource needs plus wastes, yet the average person uses 5.7 acres (0.02 km$^2$). This startling calculation makes one thing clear: People use more land than the planet holds. There may be no more compelling argument for the need to reduce waste volume.

Compaction works on either nonhazardous or hazardous noncombustible wastes. Compaction equipment companies offer machines that crush MSW into smaller, denser structures and other machines designed for low-level radioactive wastes (LLRW). As a result, a smaller overall waste

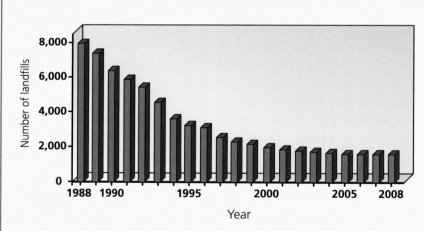

**Number of Landfills in the United States 1988–2008**

© Infobase Publishing

The number of operating landfills in the United States has been decreasing, either because they are full or because of safety hazards. The environment group Zero Waste America estimates that more than 10,000 old, unused landfills remain. By 2010, new landfills opened as old ones closed, and many existing landfills are now increasing their capacity.

volume conserves landfill space and extends the years in which landfills have room to accept new loads.

Before waste can be compacted in machines, workers first remove toxic materials. Even after dangerous components such as solvents, preservatives, and metals have been removed, some wastes are simply too bulky for landfills to accept. White goods, for example, require extra treatment for reducing their size. Salvage companies possess specialized tools and torches to cut up refrigerators, ovens, washers, and dryers into small pieces, a procedure called *sizing*. If the pieces need even more reduction before going to a landfill, the salvager shreds them. Shredding is any method of shearing, tearing, or chopping to reduce an item's size.

Compaction, sizing, shredding, and incineration all reduce waste volume. In many instances, more than one of these treatments act in sequence to transform large items into an eventual small pile of incinerator ash. Combination waste treatments have become popular in North America, Europe, and Asia, all places with regions of dense population and scarce land. Compaction alone or in combination with other treatments makes many materials easier to recycle so that they avoid the landfill altogether. This chapter examines methods of waste reduction and the tools used in waste compaction. It also covers the types of waste that are most suitable for compaction and new waste reduction technologies on the horizon.

## NONHAZARDOUS SOLID WASTE

Most reduction and compaction methods treat nonhazardous solid wastes, meaning MSW and LLRW. Compressors apply force to waste loads to reduce their size, but the composition of each waste load determines the extent to which the load can be compacted. Plastics compress differently than metals; soft metals such as tin compress differently than strong metals such as stainless steel. Paper has its own characteristics. Therefore, sorting the wastes is as important in compaction as it is in most other waste treatment methods.

Nonhazardous wastes come mainly from households and businesses. A small business running a single office may generate many pounds of wastepaper daily, while larger operations quickly accumulate tons of hazardous wastes. Compacting equipment has therefore been designed to

Small-scale compactors reduce the volume of wastes produced by households, schools, and business offices. These compactors commonly use a crushing container, pictured here, or a vise-style crusher to compact dry, nonhazardous wastes. *(Rotorpac)*

handle various waste volumes and different materials. Compacting machinery has models to treat construction debris (brick, concrete, wood beams, pipe, and wire), paper and corrugated boxes, cans, restaurant waste, or mixed wastes. Machines are even capable of compacting polystyrene packaging material and expanded polystyrene, which is solid foam similar to Styrofoam used in construction and large item packaging.

Compacting companies have separate equipment for LLRW. These wastes usually contain small amounts of radioactive materials dispersed throughout large amounts of inert matter. Though LLRW is rarely a health hazard, it is like any radioactive substance in that it requires careful attention to its waste stream. Some LLRW compactors are used at the origin of the waste; these operate most often at nuclear reactor sites, hospitals, and university laboratories.

## VOLUME REDUCTION

Crushing an empty soda can by hand is an example of waste compaction. A 12-ounce can changes volume from 22 cubic inches (360 cm³) to less than five (82 cm³) in about two seconds. Beginning in the 1970s, many new homes had garbage compactors in their kitchens to take advantage of this simple principle. Not only were total waste volumes reduced, homeowners saved money on garbage pickups.

Garbage trucks play an underrated role in the volume reduction of wastes; these vehicles usually provide the very first of many compaction steps in waste management. *Waste collection vehicles* (the term used in waste management) came into use in the 1920s at about the same time horse-drawn vehicles were being retired. Today the familiar garbage truck serves almost every community in the United States.

As a garbage truck rumbles through a neighborhood picking up waste loads, it compacts the collection with a pneumatic crusher. Most trucks contain crushers that reduce waste volume by a third of the original volume. If this compacting were not done, 30 percent of the tab paid to the hauling company would be for waste and 70 percent for air. Garbage compaction inside a truck's Dumpster is but a small contribution to total volume reduction, but it helps efficiency by decreasing the number of trips trucks must make to a waste collection center.

Landfill owners today are called upon to use their diminishing landfill space wisely. Richard Tedder of Florida's waste management division told an Associated Press reporter in 2006, "We've only got the size of the planet. Because of all of the pressures of development, people don't want landfills. It's going to be harder and harder to site new landfills, and it's going to be harder for existing landfills to continue to expand." Compacting helps alleviate the pressures on near-capacity landfills in two ways. First, it enhances the landfill efficiency by filling less space with a smaller and denser material. Second, compaction increases the profitability of a commercial landfill by lowering operation costs, consisting of labor, land, and equipment.

At the landfill, garbage trucks dump their partially compacted loads and heavy vehicles then roll over the new load to crush it again. Landfill compactor vehicles use a heavy roller studded with thick eight-inch (20-cm) teeth. The teeth plus the weight of the vehicle break up and compress materials as the compactor rolls over them. Some landfills include an additional crushing step with powerful stationary compactors before putting the dense waste load onto the landfill pile. Overall, the compacting equipment creates loads with densities of at least 1,200–1,400 pounds per cubic yard (712–831/m$^3$). Bulldozers then rumble onto the crushed waste and spread it into an even layer. As a final step, the bulldozers cover the pile with a layer of soil to reduce odors and pests. These soil layers also exert a moderate amount of pressure and compact the pile further as time passes.

Compaction works mainly on wastes that are not combustible so cannot be incinerated. This method may best serve communities that do not want incineration but find the latest thermal treatments too expensive. Standard compactors used by municipalities and businesses reduce waste volume by about one-third, depending on materials in the load. New supercompaction technology, however, reduces volume to one-fifth the original

Landfill compacting involves spreading the wastes evenly over the landfill pit and then rolling over the pile with heavy equipment that crushes, flattens, and condenses the waste. This process helps landfills conserve space so they can accept additional waste loads. *(Jeff Breedlove)*

size or less. Eventually, even the strongest compactors reach a physical limit of how much any material can be compressed; compaction greatly reduces waste volume but it does not eliminate the waste. In materials science, formulas exist to estimate the forces needed to crush different materials and the maximum degree to which those materials compress.

Waste compaction offers the advantages of being easy, requiring little effort to set up, and costing less than advanced thermal treatments. Compaction does not solve a waste problem, but it is useful as one of many tools in waste management. Later, this chapter provides examples of how dense, compacted materials also create new choices in sustainable waste management.

## COMPOSTING

Composting reduces waste volume and requires even less technology than heavy rollers trundling over landfills. Composting is the biological degradation of wastes composed of organic compounds. The final product is called compost, an organic mixture that cannot undergo further breakdown. Soil bacteria and fungi carry out most of the composting reactions in household and larger scale composting operations. These microbes require an ample supply of oxygen for their metabolism, so compost owners routinely mix the pile to increase the surfaces exposed to the air. In compost piles, exposure to light rather than microbes destroys a small percentage of the wastes in a process called *photodegradation.* Photodegradation and microbial reactions work together to decompose waste.

For many years, composting belonged solely to industrious homeowners who used it to recycle nonmeat food scraps and yard waste. Composting methods have changed little over the decades. Homeowners construct either a stationary pile needing a mix now and then, or they put their wastes into a container called a composter, which makes mixing easier. After several weeks, the compost turns into a fertilizer for the garden.

Household composters require little more than a steady input of wastes and a means of aerating the pile. Electric units mix the compost with the flip of a switch; other composters contain a handle for manually rotating the barrel to aerate the contents. Compost cones are the simplest of all composters. The cone-shaped vessel placed in a garden receives wastes and releases compost directly into the underlying soil. The owner merely moves the cone every few months to a new spot.

In addition to aeration, moisture level influences the biological reactions inside compost piles. Compost with high carbon-to-nitrogen ratios—leaves are an example—tend to have low moisture and therefore take a long time to degrade. Wastes with low carbon-to-nitrogen ratios and high moisture, such as grass clippings, degrade faster.

People building sustainable communities have turned to composting for efficiently managing household waste with almost no expenditures of money or energy. Cities in Austria, Denmark, Germany, Switzerland, and other European countries now compost more than 85 percent of their yard trimmings, paper, and vegetable food wastes. By comparison, U.S. communities compost only 5 percent of their organic wastes, most of it yard waste. By throwing away most of their household garbage, families also discard a valuable ingredient for recycling Earth's nutrients and sparing landfills. The success of composting may rest on the ability of people to see garbage in a different light. "This rethinking can be as simple as seeing household garbage as too 'valuable' to throw away, especially when we are also throwing away the health and opportunity of future generations," writes Darci Clark for a composting advocacy group (the Marquis Project) in Manitoba, Canada. "Incorporating daily/weekly/seasonal routines like diverting organic waste for backyard composting mean little inconvenience for us, but these behavior changes have a significant ripple effect on our communities and the larger world around us."

A backyard compost may be an open-air waste pile or a contained compost unit, pictured here. Composters require a regular influx of organic wastes, a source of digestive enzymes, moisture, and a temperature range optimal for enzyme activity. Specialized composters are available to act on kitchen wastes, yard waste, sewage, or pet waste. *(District of Mission, B.C., Canada)*

The organic food and slow food movements use composting to help build sustainability and aid nutrient recycling. Organic vegetables contain no pesticides or chemical additives, so compost made from them serves the next organic growing season. Slow food is a more general concept: the practice of buying locally grown foods for meals prepared at home and emphasizing good nutrition. One of the slow food movement's goals is continual recycling of the Earth's nutrients, and composting is one of the best ways to do this.

Homes designed for sustainable living use composting toilets to reduce water use and minimize the total wastewater they produce. (Some municipalities do not permit them because of safety concerns related to infectious microbes. Modern composting units inactivate dangerous microbes so they do not pose a health threat.) Though composting toilets have been used for generations, they were first sold commercially only since the 1960s in Scandinavia. As the sustainable living movement slowly spread in Europe in the following decade, builders included composting toilets in many new home plans. The sustainability movement progressed westward to North America in the 1980s and with it came a desire for low-maintenance home products with superior efficiency. New models of composting toilets were therefore redesigned to include more conveniences.

The newest models of composting toilets can serve a typical single-family home and reduce its wastes to a nonhazardous material in a few weeks. The compost turns into garden fertilizer after several months of biological breakdown. Conveniences that have been added to these models include fans for reducing odors and vacuum-assisted flushing. Most toilet companies also offer nutrient formulas to enhance microbial reactions in the composting process; some designs help further by regulating temperature and aeration. Composting toilets can be electric, battery powered, or manually operated.

Compost companies in the United States and Europe provide material on a large scale as topsoil, landfill cover, or organic fertilizers, discussed in detail in the following sidebar "Organic Fertilizers." The quality of industrially produced compost varies for two reasons: (1) different starting materials from batch to batch and (2) different compost-making devices. One thing does remain standard in commercial composting: It involves strict attention to factors that increase microbial activity. These factors are nutrient balance, particle size, moisture content, temperature, and aeration. When commercial composters pay attention to these factors, the end result is quality compost.

## COMPACTION SYSTEMS

Compaction equipment handles small waste loads of less than 100 pounds (45 kg) to large volumes of industrial wastes. Compaction is particularly important in certain industries that produce large volumes of waste each day. Industrial compactors belong to three main subclasses: balers, shredders, and a combination of the two called hoggers.

Balers, also known as waste presses, produce daily either a single bale weighing 1,000 pounds (450 kg) or 20 or more smaller bales. Most serve industries compacting paper and corrugated boxes. These models have adjustable speeds and produce bales of varying density according to the paper grade. The large balers also tie together each bale for easier loading onto trucks. Newer balers have been designed to compact the following non-paper materials: foam rubber, Styrofoam, cans, burlap bags, buckets, mattress filling, denim scrap, vinyl sheet scraps, foil, medical waste, fibers, and some food waste. A few unique balers even crush computer keyboards.

Shredders use stout blades to reduce paper waste volume. Wastepaper usually enters a shredder on a conveyer belt and then falls into a unit called

High-speed waste balers have been developed to reduce the volume of almost every type of waste, including bulky metal items and low-level radioactive materials. Large balers located at landfills increase MSW density and decrease the landfill's costs for heavy compacting equipment. In many cases, compacted material can be reused as support material for construction or as insulation. (ECVN.com)

## ORGANIC FERTILIZERS

rganic fertilizers originate from plant or animal sources raised free of pesticides, antibiotics, and other chemical additives in soil or in feed. Farms raising organic vegetables, fruits, or meat animals must use organic fertilizers if they wish to retain their organic certification. Two types of organic fertilizer come from two sources. First, green manure consists of crops grown by organic methods and then plowed back into the earth. Second, brown manure is waste produced by organically raised livestock. Both forms of manure restore the chemical balance of soil by returning essential nutrients to it.

Fertilizers used in large-scale agriculture have traditionally been synthetic substances that have been formulated to correct the nutrient deficiencies in different types of soil. Organic fertilizers often supply lower amounts of nutrients than synthetic fertilizers, but they offer the advantage of releasing their nutrients slowly into the soil. This slow release reduces the chance of water pollution from fertilizer runoff.

Regardless of a fertilizer's source, it must supply three essential nutrients, nicknamed the N-P-K formula. N represents the fertilizer's nitrogen content by percent and P is the phosphorus content. The nitrogen may

*(continues)*

Fertilizer production improves sustainability because it reduces waste volume, converts waste into a reusable product, and acts to recirculate the Earth's nutrients. Garden vegetables or agricultural crops then reabsorb these nutrients to begin a new cycle. Fertilizer supplies mainly nitrogen, phosphorus, and potassium; the fertilizer pictured here contains 10 percent nitrogen, 2 percent phosphorus, and 6 percent potassium compounds. *(Woodstream Corporation)*

*(continued)*

be in the form of nitrate ($NO_3^-$) salts such as sodium nitrate, urea (($NH_2$)$_2$CO), or as liquid nitrogen. Nitrogen in organic fertilizers comes mainly from urea. Phosphorus is in the form of phosphate ($P_2O_5$). The K represents potassium as potash ($K_2O$). A fertilizer with N-P-K of 8-30-15 contains 8 percent nitrogen compounds, 30 percent phosphates, and 15 percent potash. In addition to N-P-K, fertilizer usually contains small amounts of other nutrients such as manganese, copper, cobalt, iron, and zinc.

Organic fertilizers are used on lawns and in greenhouses in addition to organic farms. Today's brown organic fertilizers derive from manure, sewage *biosolids*, blood or bone meal, feather meal, and fish meal. Green organic fertilizers usually come from cottonseed meal, alfalfa meal, or corn gluten meal. (Meal is a coarsely ground by-product of animal or plant manufacturing.) All naturally produced fertilizers provide an excellent example of sustainable uses of the Earth's nutrients.

a shredding drum. These machines process up to 20 tons (18 metric tons) per hour, and specialized shredders handle books (soft- and hardcover) and magazines in addition to sheet paper. A sweep shredder operates in facilities that produce large or heavy waste items. Facilities position their stationary sweep shredder at a place near where waste accumulates, for instance, at one end of a production plant's floor. Workers sweep the waste into the machine that can be connected directly to a baler. In the printing industry, such shredder-balers are hoggers, which play an essential role in reducing the large waste volumes produced in printing and in destroying discarded phone books.

A more powerful compaction technology has been in use since 1978 but has recently gained renewed attention. It is called *supercompaction,* which achieves crushing forces of up to 1,500 tons (1,361 metric tons) across a waste load's surface. Supercompactors reduce waste to smaller than one-fifth of its original volume and produce a small end product called a puck, which has a density of 50–100 pounds per cubic foot (802–1,601 kg/m³). Paper and foams produce less dense pucks; metals produce denser pucks. Supercompaction compresses waste and also the metal drums and barrels that store the waste, all in one step. Trucks carry the puck to final disposal sites, such as landfills, burial sites, or incinerators.

Nuclear facilities now use supercompaction for reducing their volume of LLRW because these wastes cost more to dispose of than most other wastes. By reducing the LLRW volume, companies can reduce their overall costs. Hospitals and universities have also turned to supercompaction for handling their LLRW. Supercompaction enables these facilities to make room for new incoming loads of radioactive waste. Though compaction does not solve the overall challenge of finding space for waste, it forestalls the point in time in which all the disposal space for nonhazardous waste, especially LLRW, has been filled.

Commercial supercompaction companies operate as part of the waste management industry in Canada, Great Britain, France, the Czech Republic, Slovakia, Ukraine, Russia, Finland, China, and Brazil. Some of these places use supercompaction technology to treat air filters, asbestos, incinerator ash, concrete, and insulation, and advanced supercompaction technology is under way to compress combustible solids.

Modern waste management must overcome supercompaction's short list of limitations. First, most existing compaction facilities do not compact combustible solids, and they cannot treat liquids. Second, large pieces of metal or plastic must be sized before compaction so they do not jam the equipment. Any extra step in waste management corresponds to a cost increase. Third, supercompaction requires a considerable amount of maintenance, which adds cost.

## PAPER COMPACTION

Paper totals about 35 percent of the weight of MSW produced in the United States and is the single largest component of MSW. Despite efforts to reduce the mountain of paper waste, this waste category is rising in the United States and other countries by as much as 20 percent annually. Waste recycling programs in business offices emphasize paper, yet each year in Europe alone more than 100 million tons (91 million metric tons) end up in landfills or go to incinerators. Beginning in the 1970s, the U.S. economy transitioned into primarily a service-oriented economy rather than a manufacturing economy. The change resulted in the production of more than 80 million tons of wastepaper and cardboard per year—the United States also consumes the most paper of any country in the world.

Perhaps the term *paper economy* more accurately describes the U.S. situation today, illustrated by the following points:

- A year of daily newspapers produces 550 pounds (250 kg) of wastepaper.
- The United States uses almost 700 pounds (317 kg) of paper per person per year.
- Worldwide per person usage averages only 124 pounds (56 kg) per year.
- The United States recycles little more than 50 percent of its wastepaper.

*Environmental Issues,* an online newsletter, explained in 2008, "Why is paper recycling such a challenge? The answers have to do with the natural reluctance of people to change habits, with the designed-to-fail nature

Paper constitutes a large portion of waste in industrialized countries, and it also presents one of the biggest compacting challenges. Builders of new sustainable homes experiment with compacted paper as insulation material. *(Cheryl Graham)*

of many programs, and the assumption of managers that such programs will run themselves. As a rule of thumb, a typical office generates about 1.5 pounds [0.7 kg] of waste paper per employee each workday. (Financial businesses generate more than 2 lbs.) That's roughly 350 pounds [159 kg] per employee per year—or a total of about 2.5 tons [2.3 metric tons] for a small fifteen-person office."

More newspapers get recycled than other paper products, yet tons of newspapers still go to landfills each week. Compactor vehicles have a hard time compressing landfill loads high in paper, so waste-handling companies usually rely on balers to first turn newspaper and other paper into dense bundles. A typical paper baler reduces the volume of loose cardboard by 85–90 percent within 30 seconds. Smaller office units reduce scrap paper volume by almost 95 percent and then automatically tie each bale with wire. The smaller, denser bales help minimize trash pickup fees and reduce fire hazards.

Compaction represents the first step in paper recycling. After wastepaper arrives at a recycling facility, the recycling company removes the inks from the pulp, then bleaches, rolls, and dries each batch. Paper manufacturers then produce the smooth flat sheets familiar to consumers by sending the treated pulp through a series of additional compactions. Paper recycling therefore advances sustainability by conserving natural resources. Recycled paper has also been adopted to many new products.

## PRODUCTS FROM COMPACTED WASTE

Green architecture and construction in sustainable communities have put a premium on raw materials from recycled wastes. At present, almost all of the materials familiar to households have been tried in some sort of recycling as long as they do not cause health hazards in humans or the environment.

Building construction, road building, and highway support structures together make up the largest market for compacted wastes today. Compacted pucks and bales provide strength to structures, and they can bear heavy weight. Construction companies often add more strength to bales and pucks by mixing in concrete or mortar. These additives fill the large spaces and small pores within the compacted material and therefore further increase weight-bearing capacity.

Other companies specialize in making raw materials out of wood wastes and salvaged or *reclaimed wood* has now become a fast-growing

Massive machinery crushes cars, farm equipment, and appliances into flattened scrap to be transported to a metal recycler. Workers first strip the waste of all useful parts, plastics, and fabrics. Most new cars today come from older cars that have been crushed and recycled. *(Louisiana Department of Environmental Quality)*

business. Salesman Larry Percivalle told the *Los Angeles Times* in 2006 that wood salvaged from abandoned buildings is "wood that comes with a great story." Wood compacts into strong veneers or materials for building decks, railings, and roofing. Resurgence in log homes has also increased the market for compacted wood fibers, which are discarded by the lumber industry during lumber production. A small percentage of homeowners prefer biomass stoves that run on *biofuels* made of compacted wood pellets. Overall, recycling wood products aids in forest conservation and affects in a positive way humans' ecological footprint. Nadav Malin, editor for *GreenSource* magazine, said in the same *Los Angeles Times* article, "If you were to try to get wood of that quality from new trees, you would be cutting old growth somewhere and perhaps affecting sensitive ecosystems."

Compressed plastic drink bottles have a long history as a reusable material. Compressed plastic goes into the making of outdoor furniture, paneling, fences, decks, and walkways. Plastic lumber provides advantages over many other types of construction materials because the dense, compacted plastic resists harsh weather, does not rot, and does not splinter. Indoors, recycled plastic serves in almost every hard item or surface in houses, offices, and classrooms. Compacted plastic, in fact, may be a better choice than new plastics for many items because of compacted plastic's greater strength and resistance to bending and denting.

Compacted wood-plastic composites provide the advantages of both components. Wood provides strength; plastics provide heat and water resistance. These composites go mostly into furniture, decking, playground equipment, and the interior moldings in cars and trucks.

One of the fascinating sights in recycling is a stack of crushed cars on a flatbed truck headed for a metal salvager. Cars represent one of the most recycled products in all of industry; almost 100 percent of salvaged junked cars supply reusable metal. The Environmental Defense Fund reports, "Most cars are recycled at junk yards for profit, and the business accounts for one-third of U.S. steel scrap. If your car 'dies,' make sure it ends up with an auto wrecker, who will strip it of reusable parts and sell them. What remains of the car is shredded and separated into ferrous metals (iron and steel), non-ferrous metals (zinc, aluminum, copper, and brass) and 'fluff' (seat cushions, carpeting, dashboard, various plastics, etc.). The metals are sold to be remelted in this country or abroad. So far, there is no market for the 'fluff,' which is sent to landfills." The recycling industry, however, has been making greater use of the excess materials as insulation and filler for construction materials.

To prepare cars for recovery of their metals, car recyclers first remove tires, glass, fluids, and nonmetal interior parts from junked cars and then use hydraulic compactors to crush each car into a metal "pancake." Within minutes crushing machinery flattens a sedan to two feet (0.61 m) in thickness or less. Cranes then stack the crushed cars onto a truck or train and off they go for metal recovery. About two-thirds of a car's weight is steel and iron with smaller amounts of copper, lead, and aluminum. The steel industry receives the majority of vehicle scrap metal and uses it for producing new car bodies. The steel that goes into today's cars and trucks has likely been in several cars previously and should end up in many more. A new business—and even a new form of crime—has recently developed from the recycling of valuable metals inside the catalytic converters of cars and trucks. Catalytic converters contain ceramic cores coated with the metals palladium, rhodium, or platinum. Though the value of metals changes daily on world markets, thieves have discovered that removing catalytic converters from cars nets them a profit by selling the parts to recyclers who deal in precious metals. The metallurgist Hossein Arbabi of San Leandro, California, explains why legitimate businesses and thieves desire this single item: "Platinum right now is worth more than gold." In fact, platinum's value can reach twice that of gold, explaining why it is the most valuable recyclable material in a car.

# CONCLUSION

More wastes are being generated than ever before, and one problem this creates is the sheer volume of waste that must be handled, hauled, buried, or treated. Waste reduction is therefore a priority in waste management. There are two aspects to waste reduction: (1) source reduction in which less unusable materials arise so that less ultimate waste is made, and (2) physical reduction by applying pressure to a waste load to decrease its volume. This second approach is called compaction. Almost any nonhazardous solid waste can be reduced to at least one-third of its volume by compaction. Reduced waste volume helps prolong the lifetime of landfills, and the compressed materials might have value as reusable materials.

Compacted metals, woods, and plastics today serve as raw materials for manufacturing products used in construction, highway projects, and households. The density and the strength of compacted materials make them good choices for weight-bearing structures. Therefore, new furniture, construction materials, and roads usually receive some recycled waste that has been turned into a useful raw material.

Supercompaction offers an advanced type of compaction that applies higher than standard compaction pressures to wastes. It results in a smaller, denser bale or puck, which acts as construction material.

Composting is a form of waste reduction in which microbes digest organic wastes and turn them into fertilizer. Composting has moved beyond backyard compost heaps: Specialized companies now make organic composts from chemical-free wastes. Composting toilets have also started to gain acceptance in new sustainable buildings.

Compaction reduces the volume of waste, but it does not eliminate the waste or chemically convert it to another form. Compaction does, however, play a role in the overall management of municipal solid wastes and low-level radioactive wastes. Compaction does not require complex technology, and it costs less than advanced waste treatment methods. This waste treatment approach also works well in communities not able or willing to use incineration or newer high-efficiency thermal methods.

# WASTEWATER TREATMENT

Wastewater treatment is the process of turning contaminated waters into clean, reusable water. Hazardous materials in wastewater originate mostly from sewage, which is toilet and drain waters collected into sewer systems. For this reason, wastewater treatment is often referred to as sewage treatment even though waters in addition to sewage flow into treatment plants. City storm drains collect irrigation runoff, rinse waters from homes and businesses, and heavy rains, all of which contain a heterogeneous mixture of soluble and insoluble substances. In addition to sewage, wastewater contains the following hazardous constituents: oils, gasoline, paints, solvents, detergents, animal waste, pesticides, fertilizers, acid rain, and industrial chemicals. Nonhazardous materials usually found in wastewater are leaves, lawn trimmings, gravel, sand, and other materials that are not chemical or infectious. Wastewater treatment plants have no way of knowing the exact composition of the thousands of gallons of material arriving each day, and, for that reason, untreated wastewater must be thought of as hazardous waste regardless of its source.

This chapter covers physical, chemical, and biological steps that combine to treat wastewater and turn it into an end product that does not harm the environment. The chapter also reviews the interesting history of wastewater treatment and new technologies for making wastewater treatment more efficient and sustainable.

## THE HISTORY OF SANITATION

Evidence of wastewater treatment dates to 1700–1500 B.C.E. in Knossos, the capital of the Greek island of Crete, where plumbing and waste removal systems served the higher classes. Later, the Greeks developed a form of today's municipal landfill and city-run waste pickups. Citizens of the Roman Empire enjoyed advanced water and sewer distribution systems from about 100 B.C.E. to the end of the sixth century. These latrines—their so-called rooms of easement—are considered the first true introduction of sewer systems throughout present-day Europe. Europe in the Middle Ages assigned less importance to sanitary handling of wastewaters than their Roman predecessors. The people paid for this decision by suffering through plague epidemics that so devastated the population—every third person died—that the void in leadership and science affected generations for centuries. Though poor sanitation led directly to the spread of the plague or the Black Death, not until the 1800s did a London doctor named John Snow make a connection between disease and waterborne bacteria. In 1849, Snow summarized a detailed study he had made of cholera outbreaks and contaminated water in a pamphlet named *On the Mode of Communication of Cholera.* In Snow's summary section he noted, "Care should be taken that the water employed for drinking and preparing food (whether it come from a pump-well, or be conveyed in pipes) is not contaminated with the contents of cesspools, house-drains, or sewers . . ."

During the time Snow's writings were published, England wrote its first public health code and launched what came to be known as "the age of sanitation." The English began building infrastructure for carrying sewage, and public officials monitored the sewer lines for leaks. Perhaps part of their newly discovered fervor for sanitation came about after Snow's pointed description of their health problem: ". . . the slops of dirty water, poured down by the inhabitants into a channel in front of the houses, got into the well from which they obtained their water . . ."

In 1895, New York City set up this nation's first system for managing public garbage, including the transport of municipal sewage. Public health officials soon set up treatment plants in which sewage would filter through fine sand beds to clean out the largest pieces of contamination. These early sewer systems had a number of flaws, however, and often contained gaps and breaks in the pipes that allowed wastes to disappear into the environment. Meanwhile, smaller towns struggled with even less adequate wastewater treatment. Though water distribution and wastewater transport have improved in the United States, today many areas in the country do not have access to sewer lines and rely on *septic systems.* In other parts of the world, wastewaters flow uncontrolled into the environment as they did thousands of years ago. Wastewater treatment is therefore an area of waste management in perhaps the greatest need of advances.

## CONSTITUENTS OF WASTEWATER

Municipal sewage contains the liquids and semisolid wastes produced by communities. Domestic sewage originates solely from households, and industrial sewage comes from businesses and large industrial centers. The water entering a wastewater treatment plant has physical, chemical, and biological components that often relate to each other and affect the water's treatment. The major components of wastewater treatment plant inflow are the following:

- suspended solids
- biodegradable organic compounds
- persistent or nonbiodegradable organic compounds
- pathogens
- pollutants known to cause cancer, mutations, damage to developing fetuses, or toxicity
- heavy metals
- dissolved inorganic calcium, sodium, and sulfate compounds
- nitrogen, phosphorus, and carbon compounds
- antibiotics and other drugs

Hazardous constituents usually originate from biological or chemical sources. Biological hazards in wastewater consist of pathogenic microbes such as viruses, bacteria, and protozoa plus larger organisms such as helminth (worm) eggs and larvae. These hazardous things come from homes and public buildings through sewage collection systems and in runoff from farms and open spaces. Chemical hazards consist of pesticides, detergents, metals, and various drugs, hormones, solvents, fuels, and other organic compounds. Physical constituents such as sand and dirt are not hazardous, but they may have hazardous pathogens or chemicals attached to them. The wastewater treatment plant takes on the responsibility of reducing the amount of all of these biological, chemical, and physical constituents to safe levels before the treated water returns to the environment.

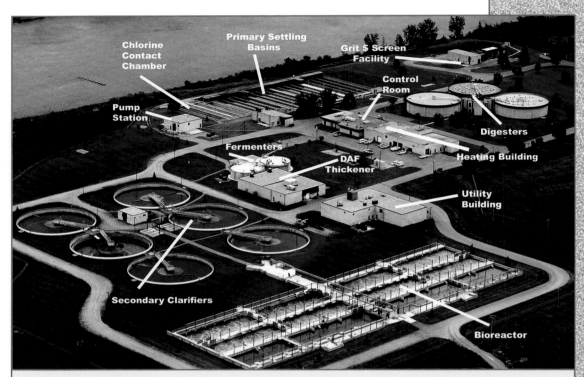

This wastewater treatment plant in Saskatoon, Canada, uses a combination of biological treatment (bioreactors, digesters, fermenters), physical treatment (screens, clarifiers, settling tanks), and chemical treatment (chlorine chamber). Organic matter is decomposed in the fermenters and digesters without oxygen; it decomposes in the bioreactor in the presence of oxygen. The dissolved air flotation (DAF) thickener is where air thickens the solids so they can be scraped off the liquid. *(Saskatoon, Canada)*

The Environmental Protection Agency's (EPA) Office of Wastewater Management sets the upper limits allowable for each constituent in treated wastewater outflow, also called effluent. Laboratories at wastewater treatment plants contain equipment that analyzes the levels of each of these constituents in the influent, the water entering the plant, and again when the treated water is about to exit the plant. The difference between influent level of any constituent and its effluent level equals the amount removed during the treatment process, as follows:

wastewater influent − wastewater effluent = amount of wastes removed

Treatment plants analyze the physical characteristics of wastewater first because many physical characteristics of water determine how easy or difficult it will be to remove contaminants. In general, the more dirt called solids present in wastewater, the more effort will be required to clean it. Technicians use more than one technique to measure wastewater solids; dissolved solids, suspended solids, particle size, and turbidity (cloudiness) are all names for different types of insoluble particles that must be measured in wastewater. In addition technicians measure the color, odor, density, temperature, and conductivity (ability to conduct an electrical current) of the wastewater after it has been treated as clues to whether the water is safe for the environment.

Biological testing makes up the next important step because wastewater carries high levels of hazardous biological matter. Technicians use microbiology methods to measure the levels of bacteria in the water and thus determine the overall safety of wastewater as it goes through a series of treatment steps. The laboratory staff also monitors the concentration (as number per mL) of protozoa, cysts, eggs, and larvae by counting each of these constituents under a microscope.

Biological oxygen demand (BOD) is a measurement that has long been important in wastewater analysis and is used in no other science. BOD indicates the amount of oxygen available for microbes to use in their waste degradation activities. Said another way, BOD relates to the amount of organic compounds in the water and is another measurement of the cleanliness of treated water. The BOD technique takes five days before a technician receives the test's results, and this long waiting period is the greatest drawback of the BOD test. A BOD value describes water that left the treatment plant five days before! For this reason, new technologies such as probes are replacing BOD in wastewater analysis.

Nucleic acid probes provide very sensitive detections of low levels of matter in water and produce their results within a day. Probes contain radioactive molecules that bind to microbial nucleic acids, either deoxyribonucleic acid (DNA) or ribonucleic acid (RNA). The amount of radioactivity in a water sample indicates the amount of microbes in the water. Another type of indicator, a fluorescent probe, works in a similar way but gives even faster results. Instead of radioactivity, fluorescent probes contain a light-emitting molecule attached to an antibody that recognizes a certain microbe. The amount of light produced by this reaction correlates with the amount of microbes in the water. Probe technology used in wastewater analysis has also been valuable in other types of environmental analysis, mainly the determination of sewage contamination of surface waters, beaches, and soils.

Chemical analysis of wastewater includes the following measurements: pH, chloride and sulfate concentrations, metals, dissolved gases, chemical forms of nitrogen (ammonia, nitrites and nitrates, and total nitrogen compounds), and organic and inorganic phosphorus compounds. Nitrogen and phosphorus compounds have a special effect in treated water as is described in the following sidebar, "Nitrogen and Phosphorus." The chemical content of water must be monitored carefully, because effluent released into a bay or the ocean must not carry high levels of these two nutrients. If treated wastewater effluent puts too much carbon, nitrogen, or phosphorus into natural waters, eutrophication take place, and aquatic ecosystems die.

## NITROGEN AND PHOSPHORUS

Nitrogen and phosphorus are essential elements in plant and animal life. English physician Daniel Rutherford discovered nitrogen gas in 1772 by burning away the oxygen inside a jar and noticing that another gas remained. He showed also that the gas, nitrogen, could not support life, meaning it is an inert gas. In nature, nitrogen takes the form of a gas ($N_2$) or as a bound form as the central molecule of amino acids and nucleic acids. Nitrogen therefore takes part in the synthesis of all proteins, DNA, and RNA. Phosphorus, by contrast, does not exist free in nature. German chemist Hennig Brand discovered phosphorus vapor in 1669 in an experiment in which he heated urine in an attempt to destroy all its organic constituents; the phosphorus vapor glowed as it burned (the origin of the term *phosphorescence*). In nature, phosphorus forms a phosphate bond, which plays a crucial role in cellular energy metabolism and stores the energy in adenosine triphosphate (ATP). Phosphate groups act as energy storage forms in DNA and RNA in addition to ATP.

Biota cannot carry out metabolism without nitrogen and phosphorus. This explains why "starved" microbes living in water burst into a period of rapid growth, known as a bloom, when large influxes of nitrogen or phosphorus compounds enter the water. Blooms demonstrate also how even essential nutrients can disrupt ecosystems when these nutrients are at high concentrations.

# WASTEWATER TREATMENT AND DISPOSAL METHODS

Wastewater treatment plant design mimics the activities found in natural wetlands. In both treatment plants and wetlands, a slow flow rate allows sediments time to settle to the bottom and gives bacteria time to digest organic matter. In wetlands, vegetation absorbs some toxic compounds. Wastewater treatment includes a similar step that removes toxins by capturing them within chemical aggregates or flocs. In wetlands, natural plants and their roots filter out a portion of water's solids; wastewater treatment plants also pass the water through filters for the same purpose.

Towns locate treatment plants in low-lying areas to collect runoff from higher elevations. Treatment begins when wastewater enters the facility at a point called the headworks where the primary treatment of the influent begins. (Large industries perform pretreatment on their own wastewater by removing most or all of the pollutants before discharging the water into the environment.) Primary treatment in the headworks involves the removal of large insoluble debris as the water passes through one or more large screens. The water then moves to a grit chamber where large particles quickly settle to the bottom and suspended materials stay in the liquid. The water, heavy with organic matter, moves to a series of tanks for *clarification,* which is a term used for the slow settling of small particles.

The next step, secondary treatment, takes place in a series of tanks that gradually turn wastewater into clear water. Here, bacteria that have been added to the mix digest dissolved solids and organic matter. Wastewater treatment therefore resembles bioremediation because specific bacteria degrade wastes with very little help from humans. Treatment plant workers do provide some help to the bacteria, however. This help comes from air that machines bubble through the bacteria-liquid mix to assist bacterial growth and waste breakdown. This supply of air to the digestive bacteria takes place in an aeration tank and the aerated mixture is referred to as *activated sludge.* The air serves another purpose: It prevents eutrophication in the plant's treatment tanks. Too high an influx of nutrients would lead to a harmful overgrowth of bacteria just as it does in bays and the ocean. The biodegraded liquid then passes through more filters and settling tanks that contain fine sand, which retains the tiniest remaining pieces.

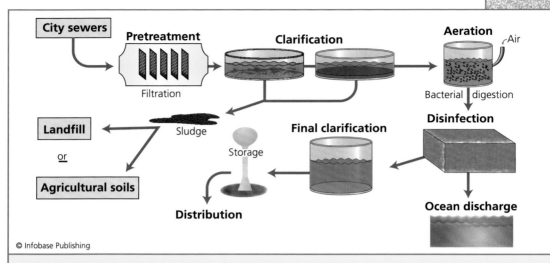

City sewers

Pretreatment

Clarification

Aeration

Air

Filtration

Bacterial digestion

Landfill

Sludge

Final clarification

Disinfection

or

Storage

Agricultural soils

Distribution

Ocean discharge

© Infobase Publishing

Modern wastewater treatment removes health hazards from wastewater so that it may be released into the environment or even used to supplement drinking water sources.

Tertiary treatment begins when chemicals withdraw the last amounts of organic compounds, and it ends with disinfection. Chlorine compounds serve as the most effective disinfectants, but they have a disadvantage: A small amount of chlorine may react with organic compounds to form dioxins or other chlorinated pollutants. To avoid this potential problem, some cities have converted to ozone or ultraviolet light to kill the microbes in the water before it is released into the environment. The decontaminated effluent is similar to *gray water* produced by homes and businesses, meaning the water is safe for the environment but not suitable as drinking water without further treatment.

The filtration, settling, and clarifying steps each produce a layer of sludge left over after the wastewater passes through each tank. This sludge is called biosolids or biomass. Biosolids may be diverted to agriculture or forest land as a nutrient addition to soils for better crop or tree yields. Alternatively, they can be compacted, dried, and used as landscaping material, in road construction, or as covering for excavated cleanup sites and landfills. Some portion of biosolids may also stay at the treatment plant where pumps push them into an enclosed tank that contains anaerobic bacteria. The bacteria continue digesting the sludge and produce methane gas during their degradation reactions. This methane can be used by the facility to supply all or part of its energy needs. When treatment plants capture methane and convert it to energy production, the facility is called a waste-to-energy (WTE) plant, and the methane represents *bioenergy* because it comes from a biological source. WTE wastewater treatment plants will undoubtedly play an important role in new sustainable communities.

People who work with biosolids follow local and state rules on where and how they distribute the material. Biosolids cannot be applied to land within a certain distance of residences, surface waters, wells, or roads. This minimum distance is called a *setback*. A setback acts as a buffer zone or safety zone, and it reduces the chance that a hazardous material will come in contact with humans or animals.

Some treatment plants include oxidation ponds or sewage lagoons to add an extra cleanup step to tertiary treatment. Developing countries commonly use these low-cost and low-maintenance ponds for all or part of their wastewater treatment, but in advanced wastewater treatment they serve mainly as a supplementary process. Oxidation ponds belong to the four following categories: aerobic, aerated, anaerobic, and facultative. All provide additional time for organic matter to decompose and for pathogens to be destroyed. Shallow (3–30 feet [1–10 m]) aerobic ponds allow good exposure to sunlight, which enhances the growth of algae that can digest organic matter in a few days. Aerated ponds make use of both sunlight and air bubbling through the water from pipes below the surface to enhance microbial digestion of wastes. Aerated ponds are the most efficient type of oxidation pond, but they require extra energy for running the aeration pumps. Deep anaerobic ponds work best for degrading sludge with very high nitrogen or protein content. These ponds require many weeks for all the sludge to digest because they rely on slow-growing anaerobic bacteria to carry out the waste breakdown. Anaerobic ponds are deeper than aerobic ponds so that the anaerobic bacteria are able to thrive at depths where oxygen is scarce. By contrast, shallow facultative ponds employ a mixture of aerobes and anaerobes, which complete digestion of the sludge in five to 30 days. Only the mechanically aerated ponds rely on energy input for waste digestion; all other types of oxidation ponds work under zero-energy natural conditions.

Septic systems operate on some of the same principles used in anaerobic oxidation ponds, except they are completely underground and not exposed to the air. This specialized method of wastewater treatment is covered in the following sidebar, "Septic Systems."

## SEPTIC SYSTEMS

Septic systems do not play a glamorous part in waste management, but they serve an essential role in treating many wastewaters. Septic systems contribute to wastewater treatment in two ways. First, they keep hazardous matter away from people and the environment, and, second, they convert hazardous materials into nonhazardous materials.

Archaeologists believe the first flush toilets were built in present-day Pakistan before 2000 B.C.E. The flush itself may have merely swept waste into a crude channel leading to a stream or river, but those first toilets provided a step toward better sanitation in the home. Water distribution and wastewater collection lines have since been linked with the growth of orderly societies by promoting health within their population. Water management also influenced where new settlements grew. By channeling drinking water to homes and using other routes to carry away wastewater, people were no longer obliged to build settlements near bodies of water. A further innovation took shape by building a sewage treatment system dedicated to a single dwelling, like a mini–wastewater treatment plant. These early septic systems did not receive rapid acceptance, however, because outhouses served almost the same purpose, and they were easier and cheaper to build. The United States, for instance, did not forsake outhouses until the mid-20th century.

A modern septic system has four parts: a pipe from the house, the tank, a *leach field*, and the soil. Collection pipes from each toilet in a house lead to the main pipe, which carries wastes to the tank. A septic tank's design allows oils and grease to float to the top of the liquid phase and solids to settle to the bottom. Screens and baffles keep the largest solids in the tank longer than the liquids so that millions upon millions of bacteria can digest the solids until the waste has somewhat clarified. In new tank designs, this breakdown occurs in 24 to 72 hours. A septic tank's daily inflow pushes the partially clarified wastewaters into pipes leading to the leach field (also called a drain

# PHYSICAL AND CHEMICAL TREATMENTS

Most of the physical steps in sewage treatment require little energy. In wastewater treatment plants, liquids flow through filters and screens without the aid of pumps, and particles settle by gravity to the bottom of large tanks. Treatment plants further conserve energy by letting the wastewater flow in a downhill direction from the headworks to the discharge end. This layout minimizes the number of energy-consuming pumps needed to keep the flow moving. In today's typical plants, less than 20 percent of energy runs the pumps and 50 to 60 percent runs the aeration tanks.

field). A leach field contains several buried pipes perforated with many small holes. Wastewater seeps through the holes into the surrounding soil where bacteria resume the digestion until the remaining organic compounds completely degrade and the water evaporates.

Contractors install fewer septic systems in the United States each year as suburbia reaches outward from large cities and sewer lines follow. Nevertheless, one in four U.S. households has a septic system. Each household is responsible for maintaining its system, a procedure that includes inspection, periodic pumping of the tank, repair, replacement of defective tanks, and monitoring leach fields. Because leaks in the collection pipe and the tank can put large amounts of infectious bacteria and viruses into the soil, as well as nitrogen and phosphorus compounds, septic components consist of materials more resistant to damage than they were in years past. Modern septic tank construction usually includes thick concrete, fiberglass, polyethylene, or metal.

Much of the efficiency of a septic system comes from a properly working leach field. Leach fields should not be located near creeks or a shallow water table, and septic systems are not recommended for cold, rainy climates, which slow the evaporation rate in the leach field. Innovations in leach field technology have addressed potential safety concerns. For instance, leach fields made of materials other than soils high in clay help the waters percolate upward and evaporate. Leach fields made of the following materials alone or mixed with soil improve the evaporation process: sand, peat, plastics, or recycled plastics. In addition, pumps and float switches reduce the chance of overflows and the contamination they cause. Septic systems aid a community's overall wastewater treatment by reducing the burden on wastewater treatment plants and the total volume of waters flowing through municipal waste streams. A properly maintained septic system serves the environment in a positive way.

Physical treatment of wastewater therefore consists of passive actions carried out through filtration and settling.

The two main chemical treatments are flocculation and disinfection, and these methods also require little energy. Flocculation takes place in one of the clarifying tanks and creates tiny clusters of particles called aggregates. These clusters form when insoluble particles come within a few micrometers of each other, allowing electrochemical attractions between them to occur. Additional charged particles added to the tank during the flocculation step augment particle aggregation. The most frequently used chemicals for aiding flocculation in this way contain alum ($Al^{3+}$), ferric chloride ($Fe^{3+}$), or polymers such as polyaluminum chloride and polyiron chloride. When aggregates grow large enough, they settle out of the liquid by gravity.

The chemical disinfection step takes place in a tank called the chlorine contact chamber. (Chlorine is the most common choice for disinfection in the United States, followed by ozone and ultraviolet light.) Chlorine disinfection kills from 98 to 99.999 percent of the pathogens in water plus the helpful bacteria that were added in earlier steps to digest organic matter. The EPA assures that treated wastewater is safe by setting an upper limit on the bacteria that can remain in the water after disinfection, and a 1973 amendment to the Clean Water Act allows wastewater treatment plants to use any type of disinfectant to meet this EPA requirement. Wastewater managers choose a treatment method according to the Best Available Technology Economically Achievable (BAT) guideline. BAT refers to selections for the best plant design and equipment as well as disinfection technology.

Chlorine disinfectant works in various chemical forms: green chlorine gas ($Cl_2$), sodium hypochlorite (HOCl) (commonly known as household bleach), chloramine compounds, or chlorine dioxide ($ClO_2$). Chlorine kills pathogenic bacteria, fungi, protozoa, and viruses before they can contaminate the environment, but chlorine itself may pose an environmental risk. Chlorine molecules react with organic compounds to form small amounts (in the ppm range) of chlorinated organic compounds. This group of diverse compounds is called *disinfection by-products* (DBP). The EPA lists more than 50 DBPs that are thought to be dangerous in the environment; many are known to cause cancer in humans. DBPs also belong to chemical subgroups: trihalomethanes, haloacetic acids, trichlorophenol, and aldehydes. Wastewater treatment plants try to treat waters as efficiently as possible in the steps leading up to disinfection for two reasons: (1) less organic matter in the water to be disinfected means less DBPs, and (2) chlorine's effectiveness as a disinfectant decreases with increasing amounts of organic matter.

Ozone and ultraviolet (UV) irradiation provide alternatives to chemical disinfection with chlorine compounds. Each method costs more than chlorine disinfection and possesses unique advantages and disadvantages. Today, countries in Europe prefer ozone over chlorine or other chemicals as a primary water treatment method, unlike the United States, which depends mainly on chlorination for wastewater and drinking water treatment. Of all of today's available methods in water cleanup, chlorination remains the most well-established technology.

Ozone ($O_3$) is a blue gas that destroys microbes in water, especially resilient protozoal cysts that resist other forms of disinfection. Ozone disinfection costs more than chlorine methods, because it requires the installation of special equipment, which partly explains why it has gained only minor use so far in the United States. Ozone may also lead to small amounts of DBPs such as aldehydes, acetic acid, and brominated compounds (similar in activity to chlorinated compounds), and these ozone

## Water Disinfection

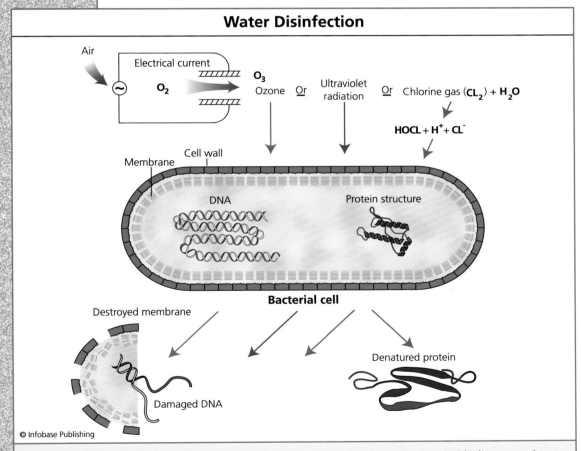

Ozone, ultraviolet irradiation, and chlorine disinfect water by destroying the mechanisms inside disease-causing microbial cells. Disinfection disrupts cellular membranes, DNA, and proteins. The World Health Organization identifies untreated or inadequately treated drinking water as a major factor in global infectious disease incidence.

by-products have been shown to cause mutations and cancer in aquatic life.

UV light at wavelengths of 100–400 nanometers damages nucleic acids and therefore halts replication in microbial cells. UV disinfection is expensive, and UV light becomes less effective in deep tanks if it cannot penetrate to the bottom. UV disinfection holds the advantage of producing no dangerous by-products. New UV disinfection technologies will soon overcome these difficulties, especially xenon lamps and eximer lamps. Xenon lamps produce pulses of UV beams over a range of wavelengths that makes the light more effective than standard UV irradiation. By contrast, eximer lamps produce monochromatic (the same wavelength) light that makes gas molecules inside the lamp unstable. As the unstable form returns to a stable form, energy in the form of photons escape, and these photons then damage molecules inside microbes. The gases most commonly used in eximer lamps are the following: xenon (Xe), xenon chloride (XeCl), krypton (Kr), or krypton chloride (KrCl).

## BIOLOGICAL TREATMENTS

Biological wastewater treatment has a longer history than either physical or chemical methods. Today's oxidation ponds and aeration tanks use biological activity to decompose organic wastes. As mentioned, this decomposition is carried out by bacteria. Researchers in wastewater treatment study different mixtures of bacteria to find the most effective blend for treatment plants.

The conversion of hazardous substances in wastewater to safe end products could not happen without microbes. Bacteria serve three important functions in wastewater treatment: (1) transformation of hazardous to nonhazardous compounds; (2) aid in forming aggregates of nondegradable particles; and (3) removal of excess nitrogen and phosphorus. Wastewater treatment uses two forms of bacterial communities to carry out these functions. The first form of community consists of free-floating cells suspended in the liquid phase. The second form of microbial community consists of complex mixtures of bacteria—in addition to some fungi and algae—called *biofilms* attached to hard, inanimate surfaces. Rock and gravel, sand, some woods, and a variety of plastics serve as surfaces for bacteria to attach to when forming a biofilm. Suspended bacteria offer the advantage of being able to digest dissolved organic compounds or very fine particles that do not settle by gravity, while biofilms have an advantage of remaining in the reaction tank for a long time so they can digest complex substances. Wastewater treatment plants tend to use both of these bacterial communities in a complementary fashion.

Wastewater microbiology involves studies on the rate at which bacteria use oxygen and nutrients and the best temperatures for their growth. Information gathered from these studies helps in the design of better wastewater treatment plants. Two of the key points in improving wastewater cleanup relate to the extent of aeration taking place in digestion tanks and the activity of the microbes in those tanks. Aeration acts with optimal flow rates and liquid temperatures to increase the efficiency of organic matter breakdown. At the same time, certain bacteria possess superior abilities for degrading hazardous chemicals such as pesticides and solvents. One goal of wastewater microbiology is to isolate superior bacteria from nature in a process called selection. Wastewater microbiologists are said to select for the best degraders of organic wastes in wastewater treatment technology. They then put these bacteria into treatment systems that have been optimized for aeration, flow, and temperature. The following sidebar, "Case Study: Wetland Waste Treatment in California" describes one community's view of wetlands for use at a municipal treatment plant.

## NATURAL TREATMENT SYSTEMS

Natural activities in soil and groundwater detoxify hazardous wastes without the need for tanks, filters, or aggregates. The natural activities in wetlands, for instance, perform all of the tasks that wastewater treatment plants perform with only one drawback: Natural processes take more time to do the same job. Wetlands have always served the environment by removing natural toxins from the environment before these substances enter aquatic ecosystems. Today, however, wetlands are the most threatened of all the planet's ecosystems. In 2005, *National Geographic* magazine described the status of wetlands in Louisiana (before Hurricane Katrina), a state that relies on a vast wetlands network for fishing and for protection from storms. The article quoted marine biologist Mark Schexnayder of Louisiana State University as saying, "Down here when we speak of wetlands loss, it's actual, physical loss. You can't stand on (the land) anymore. It's gone." Wetlands International (URL: http://www. wetlands.org/) is a nonprofit think tank that estimates that 50 percent

## CASE STUDY: WETLAND WASTE TREATMENT IN CALIFORNIA

The Arcata Marsh and Wildlife Sanctuary in northern California has since the mid-1980s combined wastewater treatment with habitat building. The sanctuary's land had once belonged to a series of local industries and a logging operation. In the 1960s, the area had outlived its usefulness for business, which closed shop but left contamination in the soil and water. The city of Arcata then turned the abandoned property into a landfill, which turned into a poorly managed menace leaking hazardous chemicals into soils and groundwaters. The festering problem might well have lasted for decades were it not for the Clean Water Act of 1972.

University engineers and the state's Coastal Conservancy teamed up to make plans for restoring a wetland that would aid the town's sewage treatment plant, a plant with a long history of dumping partially treated effluent into Humboldt Bay. Political and scientific debate soon broke out over money, manpower, and marshes; the events came to be known as the Wastewater Wars. City officials and even some environmentalists were not keen on the unique idea of ecological wastewater treatment, perhaps because they had already committed to upgrading the town's existing treatment facility to a state-of-the-art version. The expense of the new plant began to change minds, however. Then mayor Dan Hauser told the *Earth Island Journal*, "The wastewater treatment plant and pipeline would have been the largest energy consumer in the county, costing a bundle to run." Repeated scientific trials convinced Hauser and others that a wetland could and would degrade wastewater

of the world's wetlands have been lost since 1900. Human populations seem to have long since overgrown the capacity of Earth's wetlands to detoxify wastes, yet the remaining healthy wetlands continue to play a role in waste detoxification.

Natural waste treatment may also be called *intrinsic treatment* because it uses natural reactions in the environment. New sustainable communities and green homes incorporate intrinsic activities in their design, especially for wastewater treatment. Constructed wetlands have gained acceptance as a clever way to fulfill the goal of sustainable waste management. A constructed wetland is an area planted with vegetation native to the region that acts as a natural wetland to filter and detoxify wastewater. Constructed wetlands contain the following components: vegetation, wastewater distribution pipes, soils that allow water perco-

hazards to safe levels. "The regional wastewater project was a good idea in theory," Hauser said, "but it just didn't fit Arcata. When Bob Gearheart [university professor and marsh proponent] started explaining what a marsh could do and we began reading stories about what the marsh used to be like, I fell in love with the idea of marshes." In 1986, engineers began carving out a restored saltwater marsh for the purpose of treating the region's wastewater. Today, it is a wildlife sanctuary stretching more than 154 acres (0.62 km²).

Arcata's reconstructed wetlands receive pretreated wastewater from the town's treatment plant. Cattails and other aquatic plants act as sieves to filter the water and slow its flow, and a dense submerged root system absorbs excess nutrients. Algae, bacteria, and fungi—their growth kept in check by a languid flow and reduced nutrient levels—decompose organic compounds. Eventually the water flows clean and clear into Humboldt Bay. Meanwhile, the Arcata Marsh and Wildlife Sanctuary serves as a rest stop for hundreds of migrating songbirds and raptors and provides a home to a variety of shorebirds. More than 425 species of birds have been counted in the sanctuary. Arcata's marsh restoration followed a bumpy road but eventually proved that politicians and environmentalists can agree on a way to solve an ecological problem. As Mr. Gearheart concluded after the project finished, "Every group of either community or industrial people that we have worked with has wanted to do the right thing."

lation, and microorganisms. Constructed wetlands offer four advantages over today's conventional wastewater treatment plants. First, they provide habitat for birds and mammals. Second, they produce food and temporary shelter for migrating species. Third, constructed wetlands decontaminate runoff waters created in storms; these waters often bypass treatment plants during severe storms. Fourth, constructed wetlands serving new sustainable houses eliminate the need for a septic system. Constructed wetlands offer benefits even if they are not part of a sustainable community. In any town or city, a constructed wetland helps detoxify storm water, clarifies runoff from roads and driveways, and partially cleans water after hazmat spills.

Builders of constructed wetlands try to improve on the limitations of natural wetlands and swamps to speed up their reactions. For

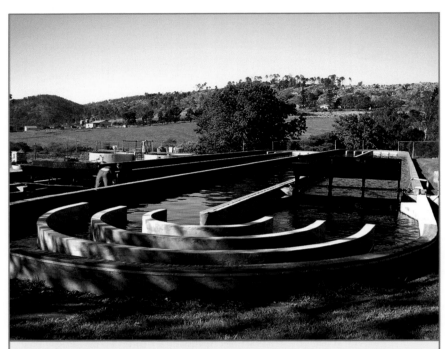

This artificial wetland in South Africa uses algae in a continuous-flow pond to further treat effluent from a municipal sewage treatment plant. *(WIO-Lab Project, United Nations Environment Programme)*

instance, certain nutrients may be in short supply in natural systems. By adding these nutrients to a wetland, natural processes continue unabated. Plans for constructed wetlands can take into account the scarcity of specific nutrients and supply them from the start. The most important nutrient to identify in wetland management is the nutrient in the shortest supply in the environment, called a growth-limiting nutrient. Nitrogen or phosphorus is often such a growth-limiting nutrient, and compounds containing these elements are easy to add into constructed wetlands from time to time. The sidebar "Nitrogen and Phosphorous" earlier in this chapter describes the significance of these elements in nature. Constructed wetlands may also benefit from the use of mechanical mixers that aerate the wetland's contents and help bacteria grow and so digest more waste. Constructed wetlands that have good nutrient supply and aeration may actually degrade wastes more efficiently than natural wetlands.

# RECLAMATION AND REUSE

Reclaimed water refers to treated or partially treated wastewater that is reusable for human activities. It is also sometimes called recycled water. Reclaiming treated wastewater helps sustain the Earth's water cycle and reduces water stress, which is the degree to which a population's water needs surpass the water supply. As sustainable manufacturing and agriculture develop in the future, their success will depend on water reclamation.

Mexico City, Mexico, was one of the first large cities to reuse partially treated wastewater in irrigation to conserve water for an expanding metropolitan population. The United States followed Mexico's lead as early as 1900. Operations known as sewage farms disposed of community wastes and at the same time recycled a portion for irrigation. Over time reclaimed

## Wastewater Recycling

© Infobase Publishing

Wastewater treatment facilities remove hazardous substances from runoff and sewage, but nature does most of the wastewater recycling on Earth. Soils, vegetation, and wetlands remove and neutralize toxic substances as they percolate downward toward groundwater reservoirs.

water became a major source used by agriculture for irrigation, the largest user, and for watering golf courses and lawns. Recycled water now supplies the following additional activities: cooling industrial machinery, cleaning vehicles, washing circuit boards, dust suppression at construction sites, soil compaction, supplementing recreational waters, supplying fish hatcheries, watering pastures, and rebuilding wetlands.

A few municipalities in the United States have applied reclaimed water in a sometimes controversial process called *recharging*. In recharging, towns pump reclaimed wastewater into aquifers that may be threatened by saltwater contamination and thus preserve their drinking water source. A small percentage of water customers in the United States use tap water drawn from aquifers that have been recharged with reclaimed water, but worries abound on the safety of drinking what is perceived as sewer water. In 2005, environmental advocate David Yetman of the University of Arizona stated to the *Tucson Weekly*, "All that's needed to stop it [wastewater reuse] is to fill beakers with water that has come out of a wastewater treatment plant and show it to people at public presentations. You can be sure that no one will want to drink it."

Few people in America and other countries may understand the technology behind wastewater treatment, so recharged water has not been accepted in many parts of the world. Brian Stewart of Australia's Urban Development Institute has admitted, "We believe this is a very sensitive issue for communities to come to grips with and we understand there are going to be concerns." Other countries with water shortages as serious as Australia's, such as Namibia, rely on recharging. The Goreangab Water Reclamation Plant in Windhoek, Namibia, was the first plant in the world to reclaim sewer water, treat it, and turn it into drinking water. In 1997, the country's secretary of water affairs Richard Fry warned, ". . . both our groundwater resources and particularly the three dams supplying the central area and specifically Windhoek are completely dry." Windhoek was tumbling toward a water calamity, so recharging attracted the town's attention. Though the city experiences customer complaints from time to time on the water's color and odor, the reclamation plant has maintained a good record in supplying safe water. Environmental microbiologist Charles Gerba of the University of Arizona has pointed out, "Reclaimed water is of better quality than most so-called 'natural' water, which we treat and deliver now." Despite the warnings from environmentalists such as David

## CASE STUDY: SAN DIEGO'S RECYCLED WATER

The 1.2 million residents of San Diego, California, experience a warm semiarid climate with less than 12 inches (30.5 cm) of rain a year. The city imports about 90 percent of its water from other parts of the state and from the Colorado River, and it must conserve all the water it can. San Diego has therefore begun a program called the Water Reuse Study to investigate all options for reclaiming used water and reusing it. The city's water department explains its problem and the reason for exploring wastewater reclamation technology: "We need to diversify our sources of water. More than 90 percent of what we use now is imported from hundreds of miles away. So we're looking at a source that's produced right here—recycled water." San Diego currently uses recycled water only for irrigation, but the Water Reuse Study intends to supplement the city's drinking water sources. The plan involves mixing reclaimed water with reservoir waters to conserve on the overall water use rate. The blend then flows from the reservoir to a disinfection chamber before it is allowed to flow to household taps.

*(continues)*

The North City Water Reclamation Plant in San Diego, California, treats up to 30 million gallons (114 million l) a day to supplement the region's water supply. These sedimentation tanks require 90 minutes for solids to sink to the bottom and grease and oils to float to the surface for removal. *(City of San Diego)*

*(continued)*

Assuring consumers of the safety of recycled water may be more of a test than treating it. Many people resist the thought of drinking water that had earlier been wastewater. San Diego had once before tried launching a water recycling program, but it met considerable opposition. The program was hurt by a number of missteps, not the least of which was its unofficial slogan, "From Toilet to Tap!" In 2007, San Diego mayor Jerry Sanders continued to oppose the reuse program even though it had been planned for more than five years. He said on a local online news service, "I'll oppose any effort to bring about toilet-to-tap," mostly based on high cost issues. Planners of the Water Reuse Study today have learned to take a different tack in convincing the public of the need for wastewater reuse technology. City Attorney Michael Aguirre countered the mayor's statement by saying, "Right now the city of San Diego is facing a water crisis. Keeping us in a system in which we are dependent on imported water from faraway sources . . . is not a prudent approach to protecting the safety and security of the people of San Diego." Water department officials now try to educate the public on the safeguards built into water treatment and water quality testing and new approaches to water conservation. The department also relies on the opinions of water reuse advocates such as Harold Bailey, director of the Padre Dam Municipal Water District, who was quoted in 2004 by the *San Diego Union-Tribune* as saying, "Virtually all of the contaminants would be gone, or they'd be reduced to a level that's lower than that required by the health department." Water officials have prudently stopped speaking about toilet-to-tap programs, but rather refer to "recycled" or "reclaimed water technology." Sometimes a bit of psychology helps convince a skeptical public of the need for new technologies.

People have a bad perception of recycled water, and yet they often fail to realize that municipal water sources, such as rivers and lakes, carry hazards. Wildlife wastes, domestic sewage, and chemicals in runoff wash into drinking water sources. U.S. water treatment is nevertheless among the safest in the world, and the same methods used in drinking water treatment can be applied to wastewater treatment technology to help in water conservation.

Yetman, microbiologists monitor reclaimed water to assure it contains no health hazards.

In at least one region of the United States, the water utility is putting concerted efforts toward using reclaimed water to meet a major portion of the potable (drinking) supply. Millions of people in arid southern California drink from aquifers recharged with reclaimed water. This is discussed in the sidebar on page 163, "Case Study: San Diego's Recycled Water." To use reclaimed water, the water utilities follow strict EPA rules. The EPA does not permit treated wastewater to go directly from a wastewater treatment plant to any human use without first spending time in another system, such as an aquifer or a reservoir. In addition, the EPA enforces stringent requirements for disinfecting potable water from reclaimed sources. Reclaimed water must be free of bacteria and viruses before it can be used and must also meet minimum chemical and physical standards for clarity, pH, BOD, and suspended solids. Water reclamation depends on the most advanced filtration methods in addition to innovations using UV light. The EPA's 2004 report *Guidelines for Water Reuse* provides information on the main uses of reclaimed water: watering in urban areas, manufacturing processes, agriculture and irrigation, landscaping, recreational uses, and groundwater recharge.

# CONCLUSION

Wastewater contains a mixture of chemical and biological constituents plus insoluble particles and dissolved solids. A portion of wastewater constituents consists of hazardous, mainly organic and inorganic chemicals and pathogenic bacteria, viruses, and protozoa. Wastewater treatment technology is therefore unique among other waste treatment technologies because it involves large amounts of infectious hazardous waste.

Wastewater treatment combines physical, chemical, and biological methods for removing hazardous substances. Physical methods include particle settling due to gravity and filtration. Chemical methods consist of aggregate formation, which aids the settling of solids and pathogens, and chemical disinfection. Biological treatment uses the activities of bacteria to digest organic matter. Both aerobic and anaerobic bacteria play a role in degrading organic compounds. Because wastewater treatment relies on bacteria for treating hazardous materials, it can be thought of as a form of bioremediation.

Anaerobic bacteria make methane gas when they degrade organic matter, and, because methane is an excellent energy source, newer wastewater treatment plants plan to use methane for all or part of their energy needs. Both WTE treatment plants and natural methods of wastewater treatment such as wetlands help in building sustainable communities. Constructed wetlands are human-built wetlands that mimic natural wetlands in the way they filter and detoxify wastes. Constructed wetlands can be made to work better than natural wetlands through the addition of nutrients and by mixing, or aerating, the waters. Constructed wetlands also contribute to water conservation and sustainability.

Reclaimed water means treated wastewater in which hazardous materials have been reduced to safe levels. Reclaimed water supplements drinking water sources in countries other than the United States. The United States, however, reserves its use of reclaimed water mainly for industrial uses, agricultural irrigation, and recharging underground sources. This recharging serves two purposes: It prevents saltwater contamination and it supplements underground water sources. New technologies in wastewater treatment and disinfection will perhaps someday make reclaimed water more acceptable as a drinking water source within the United States.

# FUTURE NEEDS

Tons upon tons of materials enter waste streams in every urban setting and in each rural area of the world, and these amounts are increasing. The future needs of waste treatment may be stated simply as Bigger, Faster, Safer. Waste management professionals must find new ways to treat the constant increase in society's waste and perform this treatment fast and in a manner that does not harm workers, neighborhoods, or animals. To alleviate a portion of the world's waste problem, waste treatment must achieve two objectives: (1) to find more efficient ways to destroy solid waste, and (2) to incorporate more ways of reusing waste materials for other purposes.

Waste treatment experts investigate the nature of different types of waste in order to devise better ways to treat it. Today wastes can be categorized according to source, composition, chemical form, or the hazards they present to the environment. Sources may be point sources or nonpoint sources. Nonpoint sources are difficult to control because they come from a multitude of small sources and may be intermittent. Composition refers to things such as metals, glass, paper, etc. The chemical forms of waste are solid, semisolid, liquid, or gas.

Like hazardous waste cleanup, waste treatment can be physical, chemical, or biological. Also like cleanup, waste management selects treatment methods based on cost and speed. Waste treatment methods range from burning or burial—methods that have been used since earliest civilization—to advanced transformations of waste into other forms of matter, advanced methods such as vitrification and plasma arc incineration. Incineration and landfilling have been put under increasing strain for the past several years because communities usually resist both of these activities. Safety innovations have improved incineration and landfilling, and both of these treatments have also shown promise for generating energy from their operations. Perhaps the biggest threat to incinerators and landfills lies in the sheer amounts of waste that wait to be handled.

New technologies in waste treatment are on the way to support the traditional treatment methods of burning or burying. Vitrification and plasma arc technology treat large volumes of waste rapidly under intense heat. Vitrification produces a solid inert glass-waste composite, and plasma arc technology reduces waste to almost nothing. These two very advanced technologies operate in few places, and they have not yet made a big impact on waste buildup. Compaction, solidification, and stabilization offer features that attack the waste problem, and they do it more inexpensively than vitrification or the plasma arc method. Compaction works by reducing the total volume of waste—it is a physical treatment. Solidification and stabilization make use of chemical interactions to immobilize waste materials and prevent them from contaminating a larger area in the environment. Biological stabilization does this on a small scale using plants and microbes. Encapsulation offers another type of immobilization in which wastes are coated with an inert material. Deep burial also serves to immobilize waste deep in the Earth's sediments.

One of the most persistent wastes today is e-waste. E-waste presents a number of problems because it is growing rapidly worldwide, it is bulky, and it contains a mixture of diverse hazardous and nonhazardous materials. E-waste does not lend itself to techniques such as burning or burying, so e-waste treatment is usually discussed in terms of dismantling, which is slow and dangerous to the people doing the work.

Wastewater treatment differs from any other type of waste treatment. It deals exclusively with liquid wastes, and it is the only waste treatment that focuses mainly on infectious microbes. Future wastewater treatment methods will be obligated to find better ways to disinfect water and to devise methods so that wastewater can be turned back into safe drinking water.

Almost every treatment discussed here holds the potential to be a waste-to-energy (WTE) operation. In the long term, WTE may be these technologies' best contribution to sustainability. The waste management industry and the U.S. Environmental Protection Agency (EPA) will need to cooperate in finding and promoting solutions to the growing dilemma of too much waste.

# Glossary

**ACTIVATED SLUDGE** suspended solids in wastewater treatment that undergo digestion by high concentrations of aerobic bacteria

**AQUIFER** underground source of water

**BIOACCUMULATION** increase in the concentration of a chemical in the tissues of organisms progressing up a food chain

**BIO-CELL** a power source in which energy is produced from biological reactions

**BIOENERGY** heat or energy derived from combusting biomass

**BIOFILM** heterogeneous layer of microbes that attach to surfaces, usually submerged in flowing liquids

**BIOFUEL** gas or liquid fuel made from biomass, usually plants

**BIOGEOCHEMICAL CYCLES** natural recycling of Earth's nutrients in various chemical forms between living and nonliving things

**BIOMASS** organic matter from plants, animal wastes, or wastewater treatment that can be used as fuel

**BIOME** a terrestrial area defined by the things living there, especially vegetation

**BIOREMEDIATION** biological degradation or neutralization of hazardous wastes in the environment; usually refers to the activities of microorganisms

**BIOSOLIDS** solid matter in wastewater and produced by humans or animals

**BLOOM** a sudden growth of microbes or plant life because of an influx of waste high in nitrogen or phosphorus

**BOROSILICATE GLASS** clear heat-resistant glass containing boron in addition to the normal glass components silica plus sodium, potassium, and calcium oxides

**BROWNFIELD** abandoned or idle industrial or commercial site where redevelopment is hampered by hazardous contamination

**CALCINATION**    use of heat to solidify wastes

**CARBON FOOTPRINT**    measure of human activities' effects on the environment in terms of greenhouse gases produced and fossil fuels consumed

**CHLOROFLUOROCARBONS** (CFC)    organic compounds containing carbon, chlorine, and fluorine that are used as refrigerants in refrigerators and air conditioners and in making plastics such as Styrofoam

**CLARIFICATION**    in wastewater treatment, the settling out of particles and aggregates from water

**COMBUSTION**    process in which oxygen combines with other molecules to form new compounds, releasing energy as heat

**COMPOST**    partially decomposed organic matter to be used as fertilizer or other types of soil conditioners

**DEEP BURIAL**    disposal method whereby wastes are put into deep sediments in the Earth's crust

**DEMANUFACTURE**    disassembly of electronic wastes for recovering reusable materials and removing hazardous components

**DISINFECTION BY-PRODUCTS**    hazardous chlorinated organic compounds formed when chlorine disinfectants react with organic matter in water or wastewater treatment

**ECOSYSTEM**    living and nonliving things that relate to each other through their activities and by their location

**ECOTOXICOLOGY**    study of the harmful effects on animal tissue caused by pollutants in the environment

**E-CYCLING**    recycling of electronic waste by dismantling and recovering any reusable parts

**EMISSIONS**    pollutants that enter the air in the form of gases, small particles, or radioactive particles

**ENCAPSULATION**    a means of stabilizing small particles of hazardous waste by coating them with inert materials, such as plastics

**EUTROPHICATION**    physical, chemical, and biological changes taking place in a body of water that has received sudden high levels of nutrients, usually nitrates and phosphates

**E-WASTE**    electronic waste; discarded electronic equipment or appliances

**FILTRATION**    method for clarifying or purifying water by passing it through a membrane containing tiny pores that retain particles of μm size

**FLY ASH**   noncombustible particulate matter formed in combustion processes, such as incineration, and released with flue gases

**FOOD CHAIN**   series of plant and animal organisms in which one feeds on the preceding one

**FOOD WEB**   network of interconnected food chains containing predators, prey, food plants, and related living things that feed on its components

**FUEL RODS**   long slender tubes that hold energy-generating nuclear material and that are inserted into a nuclear reactor core

**GARBAGE**   also trash, the nonhazardous solid wastes from households and businesses such as restaurants

**GASIFICATION**   any treatment that converts a material into a vapor, usually by using heat

**GEOMEMBRANE**   landfill liner made of porous plastic or fabric that allows gases and water to escape but holds in particles and other solids

**GRAY WATER**   treated wastewater for reuse in irrigation, industry, or release into the environment

**GROUNDWATER**   water that flows into soil and is stored in underground reserves called aquifers

**HALF-LIFE**   time required for half the amount of a radioactive element to degrade to safe levels

**HAZARDOUS WASTE**   any solid, liquid, or contained gas that can catch fire, is corrosive to skin or metals, is unstable and can explode, or can release toxic fumes or chemicals

**HAZMAT**   hazardous materials; any substance that is corrosive, reactive, explosive, ignitable, toxic, or infectious

**HEAVY METALS**   metal element that is hazardous and causes toxic effects in living tissue. Example heavy metals are arsenic, cadmium, chromium, copper, lead, mercury, and zinc

**HIGH-DENSITY POLYETHYLENE** (HDPE)   a rigid plastic that is one of the most commonly recycled of all plastics

**HIGH-LEVEL RADIOACTIVE WASTE** (HLRW)   radioactive materials from the reprocessing of spent nuclear fuels and other materials containing highly radioactive compounds

**INCINERATION**   burning process using controlled high temperatures to reduce combustible wastes to ash, water, and gas, usually carbon dioxide

**INFECTIOUS WASTE**   waste containing disease-causing substances such as sewage, feces, blood, and used medical needles, scalpels, and other contaminated instruments

**INTRINSIC TREATMENT**   method of treating waste by using natural reactions in the environment, such as decomposition of the wastes by microbes

**LANDFILLING**   procedure of disposing of hazardous or nonhazardous waste in a land disposal site so that it does not reenter the environment

**LEACHATE**   water or other liquid that has flowed through hazardous solid wastes into the soil and contains environmental contaminants

**LEACH FIELD**   part of a septic system that allows septic tank effluent to percolate into the soil and slowly evaporate

**LOW-LEVEL RADIOACTIVE WASTE** (LLRW)   radioactive materials from uranium enrichment, nuclear reactor operations, isotope production, medicine, and research; usually contains small amounts of radioactivity in large amounts of material

**MUNICIPAL SOLID WASTE** (MSW)   municipal solid waste; solid materials discarded from homes and businesses and containing mostly nonhazardous substances

**NITROGEN FIXATION**   process in which microbes remove nitrogen gas from the atmosphere and convert it into a chemical form for use by plants

**NONHAZARDOUS WASTE**   any solid or liquid waste that does not cause toxic, chemical, or infectious harm to people, animals, or the environment

**NUTRIENT CYCLING**   also biogeochemical cycling, the natural processes that recycle elements from nonliving things to living organisms and back to the nonliving environment. Examples are carbon, nitrogen, phosphorus, sulfur, oxygen, mineral, and water cycles

**PATHOGEN**   disease-causing microbe such as bacteria and viruses

**PHOTODEGRADATION**   breakdown of compounds in the environment due to exposure to sunlight

**PHYTOREMEDIATION**   removal or neutralization of hazardous wastes in the environment through the activity of plants

**PHYTOSTABILIZATION**   a means of keeping contaminants from moving through soils by arresting them on or in plant roots

**PLASMA ARC**   energy as extremely high temperature that transforms matter into plasma, which is matter in its ionized form

**POLLUTION**   physical, chemical, or biological change in air, soil, water, or food that harms human or ecosystem health

**POSTCONSUMER WASTE**    used, discarded, or leftover products and discarded packaging produced by retail stores and consumers

**PRE-CONSUMER WASTE**    waste materials, ingredients, and packaging from manufacturing products and before the products are distributed for sale

**PROTECTANTS**    compounds used on the surfaces of electronic products to minimize corrosion

**RADIOACTIVE WASTE**    wastes from nuclear power plants, weapon production, medicine, and research, or any other wastes from nuclear reactions

**RADIOISOTOPE**    an atom that emits one or more types of radioactivity: alpha particles, beta particles, or gamma-rays. A radionuclide has an unstable nucleus that emits a neutron

**RECHARGING**    method of refilling an underground water source by pumping treated wastewater into it

**RECLAIMED WOOD**    also salvaged wood, wood products recovered from abandoned buildings and reused to make new houses and furniture

**RED BAG WASTE**    medical waste containing infectious materials, usually stored in bright red-orange bags that withstand heat sterilization

**REFUSE-DERIVED FUEL INCINERATION**    method of converting waste materials into energy by burning the waste in an incinerator and capturing the heat

**RESOURCE CONSERVATION AND RECOVERY ACT** (RCRA)    the U.S. law that provides a framework for the proper management of hazardous and nonhazardous solid waste

**REVERSE LOGISTICS**    process of reusing products by returning them from consumers to manufacturers

**SALVAGING**    commercial process of recovering metals and other reusable materials from large, bulky wastes such as vehicles and appliances

**SCRUBBER**    device that removes toxic gases or particles from flue gases

**SECONDARY BARRIER**    any construction material that blocks radiation from reaching people

**SEDIMENT CYCLE**    also the rock cycle, the geologic processes that form and modify rocks in the Earth's crust

**SEPTIC SYSTEM**    collection pipes, tank, and evaporation apparatus for removing hazards from household sewage

**SETBACK**    minimum distance a hazardous waste treatment method or storage site must be from roads, homes, waterways, or other places that can be contaminated

**SEWAGE**    toilet and sink and tub drain effluent in septic systems or in wastewaters

**SIZING**    physical disassembly of large solid nonhazardous wastes by breaking them into smaller pieces for further processing

**SLUDGE**    semisolid mixture left over from wastewater treatment and containing toxic chemicals and infectious materials

**SOIL CREEP**    slow movement of soil, especially topsoil, down an incline

**SOLIDIFICATION**    chemical or physical means of converting liquid or semisolid wastes into a solid form so it will not move through the environment

**SORTING**    separation of different types of reusable wastes before recycling

**SPENT FUEL**    used nuclear material of high radioactivity recovered from energy-producing reactors or from military weapon production

**SPENT RODS**    nuclear reactor fuel that has been used to an extent that it can no longer sustain an energy-producing reaction

**STABILIZATION**    biological or chemical means of keeping contaminants from moving through soils or sediments

**STANDARD**    maximum allowable concentration of a chemical in air, water, or soil as set by the U.S. Environmental Protection Agency or other environmental agency

**SUBDUCTION**    event in which one tectonic plate in the ocean moves downward (subducts) under another plate

**SUPERCOMPACTION**    technology for compacting solid waste by applying extremely high pressure on waste loads

**SUSTAINABILITY**    ability of a system to survive for a finite period of time

**THINK TANK**    a group of experts that review current research and activities within a specific subject area for the purpose of providing advice to the government or to industry

**TOTAL EFFECTIVE DOSE EQUIVALENT** (TEDE)    sum of radiation exposure to external and internal body organs

**TOXICITY**    measure of harm a compound can do to human or animal tissue

**TRANSURANIC WASTE** (TRU)    radioactive waste consisting of elements heavier than uranium (greater than atomic number 92) on the periodic table and produced in nuclear weapons manufacture and reactor fuel assembly

**TREATMENT, STORAGE, AND DISPOSAL FACILITY** (TSDF)  a centralized collecting point for municipal solid waste and recyclable wastes

**UNIVERSAL WASTE**  subcategory of hazardous waste, containing mainly household items such as mercury-containing thermometers

**VITRIFICATION**  conversion of solids into glass; a heating process in which hazardous wastes are mixed with molten glass then cooled to form a stable and impermeable solid

**WASTE MANAGEMENT**  decisions and planning used in the reduction, collection, separation, storage, transportation, transfer, processing, treatment and disposal of wastes

**WASTE STREAM**  sources and the physical routes waste solids, liquids, or gases take to their final place of disposal

**WHITE GOODS**  in waste management, refers to large electronic appliance wastes, i.e., refrigerators, ovens, washing machines, and dryers

**WORLD HEALTH ORGANIZATION** (WHO)  an international organization that promotes cooperation between member countries in areas of human health and the factors affecting human health worldwide

**WASTE-TO-ENERGY** (WTE)  any waste treatment method that recovers end products used for generating energy

# Further Resources

## PRINT AND INTERNET

Associated Press. "Florida County Plans to Vaporize Landfill Trash." *USA Today* (9/9/06). This article describes St. Lucie County's plans to use plasma arc technology.

Balint, Kathryn. "Chemicals at Issue in Toilet-to-Tap Revival." *San Diego Union-Tribune* (1/4/04). Available online. URL: http://www.signonsandiego.com/news/metro/20040104-9999_1m4treated.html. Accessed October 7, 2008. This article covers the ongoing concerns regarding water conservation in San Diego, California.

Bernstein, Michael. "New 'Biofuel Cell' Produces Electricity from Hydrogen in Plain Air." American Chemical Society press release, (3/26/07). Available online. URL: http://www.eurekalert.org/pub_releases/2007-03/acs-nc031207.php#. Accessed October 7, 2008. This article describes a new hydrogen and enzyme–based fuel cell to generate electricity.

Blackman, W. C. *Basic Hazardous Waste Management,* 3rd ed. Boca Raton, Fla.: Lewis Publishers, CRC, 2001. A technical discussion of all aspects of waste treatment that is especially useful in explaining environmental laws related to hazardous waste cleanup.

British Broadcasting Company. "Chemical Spill Turns Rhine Red." BBC News (11/1/86). Available online. URL: http://news.bbc.co.uk/onthisday/hi/dates/stories/november/1/newsid_4679000/4679789.stm. Accessed October 7, 2008. The BBC reports on the Sandoz chemical spill.

Broder, John M. "Rule to Expand Mountaintop Coal Mining." *New York Times* (8/23/07). Available online. URL: http://www.nytimes.com/2007/08/23/us/23coal.html?_r=1&scp=1&sq=mountaintop+coal+mining&st=nyt&oref=slogin. Accessed October 7, 2008. This article covers new federal regulations that will again remove restrictions on mountaintop mining and concerns voiced by environmentalists.

Brookhaven National Laboratory. "Peconic River Remedial Alternatives: Phytostabilization." *Technology Fact Sheet.* Available online. URL: http://www.bnl.

gov/erd/Peconic/Factsheet/Phytostab.pdf. Accessed October 7, 2008. This is an informative brochure on the basics of phytostabilization.

Brown, Amanda. "Top Scientists Demand Deep Burial of Radioactive Waste." *Independent* (7/31/06). Available online. URL: http://www.independent. co.uk/environment/top-scientists-demand-deep-burial-of-radioactive-waste-410045.html. Accessed October 7, 2008. This article from Great Britain describes the advantages and uncertainties of burying the country's nuclear stockpiles.

*Business Wire.* "Computers for Families and Toshiba Launch Digital Divide Education Efforts." (7/23/02). Available online. URL: http://goliath.ecnext. com/coms2/gi_0199-1885608/Cox-Communications-Computers-for-Families.html#abstract. Accessed October 7, 2008. This online piece discusses the Santa Barbara, California, Computers for Families program, an option for reusing computer waste.

Carew, Sinead. "Consumers Still Slow to Recycle Gadgets." Reuters.com (4/17/08). Available online. URL: http://features.us.reuters.com/techlife/news/N1747 5408.html. Accessed October 7, 2008. Reuters provides an update on trends in e-cycling.

Centers for Disease Control and Prevention. *Summary of the Hanford Thyroid Disease Study Final Report* (2002). Available online. URL: http://www.cdc. gov/nceh/radiation/Hanford/htdsweb/pdf/htds_aag.pdf. Accessed October 8, 2008. This three-page brochure summarizes the federal government's 1988 study on radioactive waste emissions from the Hanford Nuclear Reservation.

Chopra, Anuj. "Developing Countries are Awash in E-waste." *San Francisco Chronicle* (3/30/07). Available online. URL: http://www.sfgate.com/cgi-bin/ article.cgi?file=/c/a/2007/03/30/MNGNNOUHQL1.DTL. Accessed October 7, 2008. This article explores the journey of e-waste from the United States to developing countries, especially India, and the associated environmental and health hazards.

———. "The E-waste Time Bomb of India." GulfNews.com (4/5/07). Available online. URL: http://archive.gulfnews.com/articles/07/04/05/10116121.html. Accessed October 7, 2008. This article describes the disturbing conditions of e-waste dismantling in India.

City Wire. "Mayor Opposes 'Toilet-To-Tap' Water Supply Proposal." 10News.com (9/13/07). Available online. URL: http://www.10news.com/news/14108898/ detail.html. Accessed October 7, 2008. A short news story on San Diego's use of reclaimed water and the strong opposition it faces.

Clare, Adrian. "E-recycling, California Style: California's Electronic Waste and Recycling Act of 2003 Went into Effect Jan. 1, but the State is Still Trying to Work

Out a Few Kinks." *Recycling Today* (4/1/05). Available online. URL: http://www.thefreelibrary.com/E-recycling,+California+style:+California's+Electronic+Waste+&...-a0132228083. Accessed October 7, 2008. This magazine article explains the benefits and drawbacks of stricter e-cycling laws.

Clark, Darci. "Composting Routines Work Well for City, Rural Families." Marquisproject.com (2003). Available online. URL: http://www.marquis-project.com/composting101/articles.html. Accessed October 7, 2008. A light and informative look at household composting and how it works.

Clarren, Rebecca. "Nuke Facility 'Downwinders' Take Energy Department to Court." *New Standard* (5/9/05). Available online. URL: http://newstandard news.net/content/index.cfm/items/1785. Accessed October 7, 2008. This article covers the heated debates over health and the nuclear wastes at Hanford Nuclear Reservation.

Connett, Ellen. "Global Warming Concerns Resurrect Incineration Debate." Interview by Amy Eddings. WYNC.org (3/29/07). Available online. URL: http://www.wnyc.org/news/articles/76417. Accessed October 7, 2008. This is an interesting interview of several environmental experts and their insights on waste, pollution, and sustainability.

Danby, Dawn, and Jeremy Faludi. "Bright Green Computers." In *Worldchanging: A User's Guide for the 21st Century.* Edited by Alex Steffen. New York: Harry N. Abrams, 2006. This chapter describes the concepts behind designing low-energy computers.

Devine, Dave, and Molly McKasson. "From the Toilet to Your Tap. *Tucson Weekly* (12/8/05). This detailed news article provides information on various community water conservation efforts.

Diamond, Edwin. "The Myth of the 'Pesticide Menace'." *Saturday Evening Post* (9/28/63). This news article illustrates the severe backlash that Rachel Carson endured in her writing career.

Eastman, Janet. "Reclaiming History, One Plank at a Time." *Los Angeles Times* (7/6/06). The *Times* investigates the business of salvaging woods and other pieces from historic homes.

Eilperin, Juliet. "Dead Electronics Going to Waste." *Washington Post* (1/21/05). Available online. URL: http://www.washingtonpost.com/wp-dyn/articles/A24672-2005Jan20.html. Accessed October 7, 2008. This short article summarizes the current state of the world e-waste crisis.

End-of-Life Vehicle Solutions Corporation 2006 Annual Report. Available online. URL: http://www.elvsolutions.org/ELVS_Annual%20Report6.pdf. Accessed October 7, 2008. This program targets the recycling of select car parts as a business decision and an environmental decision.

Environmental Defense Fund. "Carmakers Sue Over Clean Cars Law." EDF.org (12/21/07). Available online. URL: http://www.edf.org/article.cfm?contentID=4192. Accessed October 7, 2008. An eye-opening account of the efforts of U.S. carmakers to avoid California's strict antismog laws.

———. "Return Trip: How to Recycle the Family Car." EDF.org (8/15/07). Available online. URL: http://www.edf.org/article.cfm?contentID=2195. Accessed October 7, 2008. A how-to article on recovering environmentally hazardous car components and where to take them.

Environmental Literacy Council. "Incineration." Enviroliteracy.org (4/2/08). Available online. URL: http://www.enviroliteracy.org/article.php/60.html. Accessed October 7, 2008. A brief article explaining incineration with links to excellent resources.

Forbes/Wolfe Nanotech Report. "Thinking Small: Angela Belcher." (January 2003). Available online. URL: http://www.forbesinc.com/newsletters/nanotech/public/samples/nano_angela_jan2003.pdf. Accessed October 7, 2008. An interview article that provides interesting insight from a leader in nanotechnology.

Fryer, Alex. "Fate of Hanford Nuclear Waste in Flux." *Seattle Times* (12/13/04). Available online. URL: http://seattletimes.nwsource.com/html/localnews/2002117550_hanford13m.html. Accessed October 7, 2008. This news article explains the controversies regarding nuclear waste management.

Gannon, Tim. "Throwing Mud at River Cleanup." *Riverhead News Review* (2/24/00). Available online. URL: http://www.timesreview.com/_nr_html/nr02-24-00/stories/news4.htm. Accessed October 7, 2008. This article provides interesting insight on the pros and cons of using plants for cleaning up pollution.

Garber, Kent. "Technology's Morning After." *U.S. News and World Report* (12/20/07). Available online. URL: http://www.usnews.com/articles/news/2007/12/20/technologys-morning-after_print.html. Accessed October 7, 2008. This article covers the current crisis in e-waste.

Gearheart, Bob. "The Sustainable Paradigm Dance." Lecture, Humboldt State University, Arcata, Calif. (8/19/07). Available online. URL: http://sustainablesolutionsforh2o.blogspot.com/. Accessed October 7, 2008. This lecture covers sustainable methods for water conservation by a renowned expert in the field.

Geiselman, Bruce. "Groups Sue to Halt Nuclear Shipments." WasteNews.com (4/14/03). Available online. URL: http://www.wastenews.com/arcshow.html?id=03041401501&query1=Bruce+Geiselman&maxfiles=25&start=0&month=4&day=14&year=2003. Accessed October 7, 2008. This magazine article describes the conflicts that arise over shipping nuclear waste.

Greenhouse, Linda. "Supreme Court Roundup; Justices Decide Incinerator Ash is Toxic Waste." *New York Times* (5/3/94). Available online. URL: http://query. nytimes.com/gst/fullpage.html?res=990DE2DC1030F930A35756C0A9629 58260&scp=1&sq=Supreme+Court+Roundup%3B+Justices+Decide+Incine rator+Ash+is+Toxic+Waste&st=nyt. Accessed October 7, 2008. This article describes background to federal laws on toxic waste.

Halford, Bethany. "CE&N Talks with Angela Belcher: The MacArthur Fellow Talks about Tackling Science from All Sides." *Chemical and Engineering News* (4/7/08). Available online. URL: http://pubs.acs.org/cen/science/86/ 8614sci1.html. Accessed October 7, 2008. Preeminent scientists discusses the potential for harnessing biological reactions in energy-producing process.

Handwerk, Brian. "Louisiana Coast Threatened by Wetlands Loss." National Geographic News (2/9/05). Available online. URL: http://news.nationalgeo-graphic.com/news/2005/02/0209_050209_wetlands.html. Accessed October 7, 2008. An update on the severe wetlands damage and loss along the Gulf Coast.

Hanna, Mike. "Namibia's Water Shortage Threatens African Oasis." CNN.com (2/5/97). Available online. URL: http://www.cnn.com/WORLD/9702/05/ botswana/. Accessed October 7, 2008. This online article describes the disturbing effects of water shortage on African populations and animals.

Healthandenergy.com. "Smog Sends 53,000 to Hospital Each Summer." (10/6/99). Available online. URL: http://healthandenergy.com/smog.htm. Accessed October 7, 2008. This short article described the effects of smog in the eastern United States.

Keating, Rebecca. "Developers Back Water Reuse." ABC.net.au (3/1/06). Available online. URL: http://www.abc.net.au/water/stories/s1581306.htm. Accessed October 7, 2008. ABC News covers the water stress occurring in Australia.

Kreith, F., and G. Tchobanoglous. *Handbook of Solid Waste Management.* New York: McGraw-Hill Companies, 2002. This resource provides descriptions of solid waste categories, excellent definitions, and detailed explanations of decisions in waste management; the best resource in solid waste management.

Larane, Andre. "International: Ash Researchers Turn to Glass in France." *WasteAge* (8/1/97). Available online. URL: http://wasteage.com/mag/waste_ international_ash_researchers/. Accessed October 7, 2008. This magazine article provides a good description of the science of vitrification and the cost challenges.

Lear, Linda. Introduction to *Silent Spring* by Rachel Carson. Boston: Houghton Mifflin Company, 2002. First published 1962 by Houghton Mifflin. Lear's introduction describes the resistance Carson endured in bringing her science to the public's attention.

Little, Sam. "Dump It in the Mantle." *New Scientist* (5/28/05). Available online. URL: http://www.newscientist.com/backpage.ns?id=mg18625012.700. Accessed October 8, 2008. A short but clear description of deep sea burial of hazardous wastes.

Livius, Titus. *History of Rome,* vol. 1. Edited by Ernest Rhys. Translated by Rev. Canon Roberts. London: J. M. Dent and Sons, 1905. Available online at University of Virginia Library Electronic Text Center. URL: http://etext.virginia.edu/etcbin/toccernew2?id=Liv1His.sgm&images=images/modeng&data=/texts/english/modeng/parsed&tag=public&part=teiHeader. Accessed October 8, 2008. A firsthand description by ancient historian Livy on life in Rome from 750 B.C.E. to 25 C.E.

Long, Michael. "Half-life." NationalGeographic.com (July 2002). Available online. URL: http://ngm.nationalgeographic.com/ngm/0207/feature1/fulltext.html. Accessed October 8, 2008. An excellent firsthand account of radioactive waste management in the United States.

Longley, Robert. "Yucca Mountain Nuclear Waste Safe for 1 Million Years, EPA Claims." About.com. U.S. Government Info (August 2005). Available online. URL: http://usgovinfo.about.com/od/medicalnews/a/yuccamillion.htm. Accessed October 6, 2008. This short article explains the federal government's stance on long-term radioactive waste storage.

Ludwig, Udo, and Barbara Schmid. "Burning the World's Waste." *Der Spiegel* (2/21/07). Available online. URL: http://www.spiegel.de/international/spiegel/0,1518,467239,00.html. Accessed October 8, 2008. This German-published article gives an intriguing look at the debate surrounding the importation of hazardous wastes.

Maass, R. *Garbage.* New York: Henry Holt and Company, 2000. This book is a somewhat dated but enjoyable look at the waste problem, and it is an often-quoted resource in environmental science.

Magnuson, Ed. "The Poisoning of America." *Time* (9/22/80). Available online. URL: http://www.time.com/time/magazine/article/0,9171,952748,00.html. Accessed September 22, 2008. This oft-quoted article reviews the status in 1980 of waste's effect on the environment and health.

Makower, Joel. "Taking the Wrinkles Out of Paper Recycling." About.com: Environmental issues (2008). Available online. URL: http://environment.about.com/od/recycling/a/officepaper.htm. Accessed October 8, 2008. This short online article explains effective programs for recycling paper.

Margasak, Gabriel. "Nonprofit Joins St. Lucie County Effort to Bring Trash Power to Fla." TCPalm.com (3/25/08). Available online. URL: http://www.tcpalm.com/news/2008/mar/25/30gtnonprofit-joins-effort-to-bring-trash-power/. Accessed October 7, 2008. This article recounts the introduction of plasma arc technology to a Florida community.

Mattheissen, Peter. "Scientists and Thinkers: Rachel Carson." *Time* (3/29/99). Available online. URL: http://www.time.com/time/time100/scientist/profile/carson.html. Accessed April 30, 2008. *Time* presents an in-depth account of Rachel Carson's career and legacy in environmentalism.

Miller, Bob, to Bill Richardson, December 4, 1998. Available online. URL: http://www.yuccamountain.org/archive/nuctome1.htm. Accessed October 8, 2008. A letter from Nevada governor Miller to U.S. Secretary of Energy Richardson detailing the state's concerns over radioactive materials stored at Yucca Mountain.

Miller, G. T. *Environmental Science: Working with the Earth.* Belmont, Calif.: Thomson Learning, Inc., 2006. This is a general textbook on environmental science that includes an explanation of basic waste management and treatment.

Moran, Susan. "Panning E-waste for Gold." *New York Times* (5/17/06). Available online. URL: http://www.nytimes.com/2006/05/17/business/business special2/17ewaste.html?_r=1&scp=1&sq=Panning+E-waste+for+Gold&st=nyt&oref=slogin. Accessed October 8, 2008. This is an interesting account of the problems and innovations in e-waste recycling.

———. "The New Bioplastics, More Than Just Forks." *New York Times* (3/7/07). Available online. URL: http://www.nytimes.com/2007/03/07/business/businessspecial2/07plastic.html?scp=1&sq=The+new+bioplastics%2C+more+than+just+forks&st=nyt. Accessed May 5, 2008. This is an article on the technology of making biodegradable plastics.

Morrison, Richard. "What a Monstrously Wasteful Throwaway Society." *Times* (London, 2/13/07). Available online. URL: http://www.timesonline.co.uk/tol/comment/columnists/richard_morrison/article1373899.ece. Accessed October 4, 2008. This article focuses on the reasons for increasing amounts of e-waste.

National Research Council. Steering Committee on Vitrification of Radioactive Wastes. *Glass as a Waste Form and Vitrification Technology: Summary of an International Workshop.* Washington, D.C.: National Academies Press, 1996. Also available online. URL: http://books.nap.edu/catalog.php?record_id=5488. Accessed October 8, 2008. This book presents a technical overview of vitrification with an informative section on the history of this technology.

Netter, Thomas. "Mercury a Key Concern in Rhine Spill." *New York Times* (11/15/86). Available online. URL: http://select.nytimes.com/gst/abstract.html?res=F50717FD395D0C768DDDA80994DE484D81. Accessed October 8, 2008. This article provides background and consequences of the Sandoz chemical spill.

*New York Times.* "Jersey Chemical Landfill Told to Close in 10 Days." (7/9/76). Available online. URL: http://select.nytimes.com/gst/abstract.html?res=F50B 14F73F5B167493C2A8178CD85F428785F9&scp=1&sq=Jersey+chemical+lan dfill+told+to+close+in+10+days&st=p Accessed May 6, 2008. This short article provides insight into U.S. environmental issues more than 30 years ago.

——. Opinion: "Time to Deal with E-waste" (12/9/07). Available online. URL: http://www.nytimes.com/2007/12/09/opinion/nyregionopinions/CIewaste. html?scp=1&sq=Time+to+deal+with+e-waste&st=nyt. Accessed October 8, 2008. This piece discusses the New York City Council's attempts to address e-waste.

Nixon, Richard. *The President's Message to the Congress of the United States* (August 1971). Washington, D.C.: American Meteorological Society. Available online. URL: http://www.ametsoc.org/Sloan/cleanair/index.html. Accessed October 7, 2008. A fascinating insight into the politics of the day relative to the environment.

Oak Ridge National Laboratory. "Useful Metals Could Be Mined from Pond Sludge, ORNL Finds." ORNL.gov (8/9/95). Available online. URL: http://www.ornl.gov/info/press_releases/get_press_release.cfm?ReleaseNumber= mr19950811-02. Accessed October 8, 2008. This online news release describes methods being tried to recover and recycle toxic metals from a contaminated site.

O'Neil, Caitlin. "How Architectural Salvage Yards Work." Thisoldhouse.com (2007). Available online. URL: http://www.thisoldhouse.com/toh/article/ 0,,212818,00.html. Accessed October 8, 2008. This article reviews the business of recovering antique wood or metal items from 100-year-old houses.

Passer, Jerry. "Methane Credits Now Available on the Chicago Climate Exchange." *Brownfield Ag News for America* (12/6/06). Available online. URL: http://www.brownfieldnetwork.com/gestalt/go.cfm?objectid=587B8204-024E-B604-0439A56B668945FD. Accessed October 8, 2008. This short piece explains the basics of selling and buying carbon credits.

Preston, Holly H. "Bioplastics Have a Small but Growing Market." *International Herald Tribune* (11/2/07). Available online. URL: http://www.iht.com/ articles/2007/11/02/business/mplastics.php?page=1. Accessed May 5, 2008. A business-oriented article on plant-derived plastics.

Rather, John. "In Brief; Suffolk Overrides Veto on Peconic Warnings." *New York Times* (2/6/00). Available online. URL: http://query.nytimes.com/gst/ fullpage.html?res=9D0CE3DD113FF935A35751C0A9669C8B63&scp=1&s q=In+Brief%3B+Suffolk+Overrides+Veto+on+Peconic+Warnings&st=nyt. Accessed October 8, 2008. A short article on the status of river pollution in New York.

*Recycling Today.* "American Pulverizer—Shredders to Suit Any Application." (7/3/07). Available online. URL: http://www.recyclingtoday.com/. Accessed October 8, 2008. This is a marketing article that provides the features of new model waste shredders.

Rogers, Keith. "Nevada Officials Rip Yucca Mountain Document." *Las Vegas Review-Journal* (12/19/98). Available online. URL: http://www.review journal.com/lvrj_home/1998/Dec-19-Sat-1998/news/10272293.html. Accessed October 8, 2008. This news article illustrates the scathing criticism leveled at radioactive waste storage at Yucca Mountain, with opposite views from the DOE.

Rogers, Paul G. "History: The Clean Air Act of 1970." *EPA Journal* (January/February 1990). Available online. URL: http://www.epa.gov/history/topics/caa70/11.htm. Accessed October 8, 2008. Part of the EPA's detailed online history time line, an excellent description of the law and its deliverables.

Royte, E. *Garbageland: On the Secret Trail of Trash.* New York: Little, Brown and Company, 2005. Royte's book takes a philosophical view of today's garbage problem by emphasizing how consumerism leads to wastes and describing the concept of zero waste.

Satchell, Michael. "Uncle Sam's Toxic Folly." *U.S. News and World Report* (3/27/89). This is one of the early articles discussing hazardous substances associated with military bases.

Savannah River National Laboratory. "SRNL Technology Named Among 'World's Best'." Savannah River National Laboratory (5/15/07). Available online. URL: http://www.srs.gov/general/news/releases/SRNLTechnology.pdf. Accessed October 8, 2008. This press release provides a brief description of technology emerging at SRNL.

Schladweiler, Jon C. "Tracking Down the Routes of Our Sanitary Sewers." Sewerhistory.org (2004). Available online. URL: http://www.sewerhistory.org/. Accessed February 4, 2009. This article gives a surprisingly detailed history of wastewater management.

Scottish Environment Protection Agency. "SEPA Guidelines for Thermal Treatment of Municipal Waste." Available online. URL: http://www.sepa.org.uk/pdf/guidance/air/thermal_treatment_guidance.pdf. Accessed May 12, 2008. This brochure provides a helpful overview of thermal treatment methods in waste management and a short but good glossary.

Snow, John. *On the Mode of Communication of Cholera,* 2nd ed. (1854). Washington, D.C.: Delta Omega Society. Available online. URL: http://www.deltaomega.org/snowfin.pdf. Accessed October 8, 2008. This online book details accounts of Snow's work in solving the London cholera epidemic and provides an understanding of epidemiology.

Sundaram, S. K. "Vitrification: Putting the Heat on Waste." Seminar, Community Science and Technology Seminar Series, Pasco, Wash. (3/19/03). Available online. URL: http://regionaloutreach.pnl.gov/seminars/speakers/sundaram. stm. Accessed October 8, 2008. Vitrification expert discusses the science at PNNL.

Suutari, Amanda, and Gerald Marten. "Treating Wastewater with Wetlands." *Earth Island Journal* (Summer 2007): 30–31. Available online. URL: http://www. ecotippingpoints.org/resources/ETP_Vicious-Cycle.pdf. Accessed October 8, 2008. This short article recaps the wetlands controversy that occurred in Arcata, California, in the 1970s.

Tchobanoglous, G., F. L. Burton, and H. D. Stensel. *Wastewater Engineering: Treatment and Reuse,* 4th ed. New Delhi, India: Metcalf and Eddy, Tata McGraw-Hill Publishing Ltd., 2003. This lengthy resource presents an excellent description of the inner workings, chemical and biological steps, and advanced technologies of wastewater treatment.

Tetreault, Steve, and Sean Whaley. "Yucca Mountain Rail Line: Paiutes Have Terms." *Las Vegas Review-Journal* (6/3/06). Available online. URL: http://www.reviewjournal.com/lvrj_home/2006/Jun-03-Sat-2006/news/7760370. html. Accessed October 8, 2008. This articles recounts the issues related to shipping nuclear waste through the Walker River Paiute Indian Reservation.

Tetreault, Steve. "Lack of Money Spells Uncertainty for Yucca Nuke Dump, DOE Says." *Las Vegas Review-Journal* (2/19/08). Available online. URL: http://www.lvrj.com/news/15760627.html. Accessed October 8, 2008. This news article updates the controversies and ongoing debates regarding the funding of the Yucca Mountain nuclear waste repository.

*Time.* "The Cities: The Price of Optimism" (8/1/69). Available online. URL: http://www.time.com/time/magazine/article/0,9171,901182,00.html. Accessed October 8, 2008. This 1969 article gives a glimpse at the conditions of river and lake pollution at the time and the government's views on pollution.

Toto, Deanne. "Law and Disorder: California Gears Up for Implementation of the Electronic Waste Recycling Act of 2003 by its July 1 Deadline." *Recycling Today* (4/1/04). Available online. URL: http://goliath.ecnext.com/coms2/gi_ 0199-329549/Law-and-disorder-California-gears.html. Accessed October 5, 2008. This article details a new California recycling law and describes retailers' and manufacturers' responsibilities.

Trivedi, Chirag. "The Great Smog of London." BBC News (12/5/02). Available online. URL: http://news.bbc.co.uk/1/hi/england/2545759.stm. Accessed May 6, 2008. This article gives an absorbing account of London's 1952 air pollution crisis.

Turner, Wallace. "Atom-waste Blast Contaminates Ten; Atomic-waste Blast Contaminates 10 at Coast Complex." *New York Times* (8/31/76). Available online. URL: http://select.nytimes.com/gst/abstract.html?res=F00A12F7395816749 3C3AA1783D85F428785F9. Accessed October 8, 2008. A historical article describing serious hazards at the Hanford waste site.

U.S. Department of Energy. Office of Civilian Radioactive Waste Management. "Transporting Nuclear Waste" (2008). Available online. URL: http://www. ocrwm.doe.gov/transport/index.shtml. Accessed October 8, 2008. A brief update on the DOE's plans to ship radioactive waste to the Yucca Mountain waste repository.

U.S. Environmental Protection Agency. "EPA Signs MOU with the Green Grid to Enhance Energy Efficiency." April 22, 2008. Available online. URL: http:// yosemite.epa.gov/opa/admpress.nsf/8b770facf5edf6f185257359003fb69e/ 3acfc9d9c86920ca85257433005679ba!OpenDocument. Accessed October 8, 2008. This news release explains the EPA's conversion to energy-efficient computer systems throughout the agency.

———. Office of Solid Waste and Emergency Response. "Solidification/Stabilization Resource Guide," EPA/542/B-99/002. Washington, D.C., 1999. Available online. URL: http://www.epa.gov/tio/download/remed/solidstab.pdf. Accessed October 8, 2008. This booklet, though filled with government technical jargon, provides a comprehensive overview of the technology's resources and a helpful glossary.

———. Office of Special Wastes. "Crude Oil and Natural Gas Waste" (2008). Available online. URL: http://www.epa.gov/osw/nonhaz/industrial/special/oil/. Accessed October 8, 2008. This section of the EPA's Web site offers details on regulations concerning petroleum industry wastes.

———. Office of Wastewater Management. "Guidelines for Water Reuse," EPA/625/ R-04/108. Washington, D.C., 2004. Available online. URL: http://www.epa. gov/ord/NRMRL/pubs/625r04108/625r04108.pdf. Accessed October 8, 2008. This report offers a very detailed review of water reuse technology.

———. Resource Conservation Challenge. "Rubber Sidewalks Save DC's Trees and Residents' Knees" (2007). Available online. URL: http://www.rubberside walks.com/pressandmedia.asp. Accessed October 8, 2008. This EPA article discusses an entrepreneur's use of recycled rubber.

———. Solid Waste and Emergency Response. "Treatment Technologies for Site Cleanup: Annual Status Report," 12th ed. EPA/542/R-07/012. Washington, D.C., 2007. Available online. URL: http://www.clu-in.org/download/remed/ asr/12/asr12_main_body.pdf. Accessed October 8, 2008. This annual report provides an updated listing of the technologies in operation at the nation's current cleanup sites.

———. WasteWise. "Program Review: Joining the WasteWise Program" (2008). Available online. URL: http://www.epa.gov/osw/partnerships/wastewise/plan-program.htm. Accessed October 8, 2008. This is one of a series of articles explaining the WasteWise program and providing information for potential members or endorsers.

Wolfcale, Joe. "Mining Marin's Latest Precious Metal." *Marin Independent Journal* (6/27/07). This news article describes a growing crime: stealing catalytic converters for their metal parts.

## WEB SITES

California Integrated Waste Management Board. Available online. URL: http://www.ciwmb.ca.gov/Profiles/. Accessed October 8, 2008. A resource for topics in waste statistics, waste management, and environmental law.

National Library of Medicine Tox Town. Available online. URL: http://toxtown.nlm.nih.gov/. Accessed October 8, 2008. An enjoyable and interactive Web site that explains waste streams, provides hazardous chemical databases and articles on environmental health topics.

National Solid Wastes Management Association. Available online. URL: http://www.nswma.org/. Accessed October 8, 2008. This Web site is managed by a waste industry organization and is provides resources on waste transport and worker safety.

Northeast Recycling Council. Available online. URL: http://www.nerc.org/index.html. Accessed January 31, 2009. Informative resource on electronics recycling.

U.S. Department of Energy. Available online. URL: http://www.energy.gov/environment/wastemanagement.htm. Accessed October 8, 2008. The "Waste Management" pages describe nuclear wastes and law; a fairly informative resource.

U.S. Environmental Protection Agency. Available online. URL: http://www.epa.gov/. Accessed October 8, 2008. The EPA's Web site covers every aspect of pollution cleanup and treatment technology, environmental law, and new methods for reducing waste; contains a useful glossary of waste terms.

———. EcoTox Program. Available online. URL: http://cfpub.epa.gov/ecotox/. Accessed October 8, 2008. The EPA's program provides this online database in which users match any hazardous chemical in the database with selected animal species to learn the chemical's specific effect on species.

U.S. Geological Survey. Available online. URL: http://www.usgs.gov. Accessed January 31, 2009. The site provides scientific background on the fate of contaminants in the Earth.

Waste of the World. Available online. URL: http://www.thewasteoftheworld.org/. Accessed October 8, 2008. This international research group's Web site presents limited, but innovative views on the world's wastes; it includes data and resources on new treatment approaches.

Worldwatch. Available online. URL: http://www.worldwatch.org. Accessed October 8, 2008. This Web site is an excellent resource for international data on environment. Good links include *World Watch Magazine* and daily articles on international issues in ecology.

# Index

Note: Page numbers in *italic* refer to illustrations. The letter *t* indicates tables.